FACTS AND THEORIES OF AROMATICITY

FACTS AND THEORIES OF AROMATICITY

DAVID LEWIS and DAVID PETERS
Royal Holloway College
University of London

First published 1975 by
The Macmillan Press Ltd
London and Basingstoke
Associated companies in New York Dublin
Melbourne Johannesburg and Madras

SBN 333 16815 1

Typesetting and Composition by
Clairegraphics, London SW16 1BB

Printed by Thomson Litho Ltd, East Kilbride, Scotland and bound by Mansell (Bookbinders) Ltd, Witham, Essex.

Preface

This book is a critical essay on aromaticity. This subject is an old one dealt with in numerous books and innumerable papers but it seems that we still lack a genuine understanding of the subject. Indeed, the word aromatic has come to mean different things to different people and this is itself an indication that a critical appraisal of the topic is needed.

The term aromatic is a label applied to a group of molecules which seem to have something in common. It is an inexact concept for which it is probably impossible to find a rigorous definition, but it seems that scientific progress often depends on the development of broad concepts such as this, as is shown by the continuing use of the term aromatic in everyday chemistry.

The problems that have bedeviled the subject of aromaticity reflect the difficulties that have beset the subject of wave function computation. The theoretical contribution to the former subject has been a distinctly double-edged one. There can be no doubt that theoretical ideas have given great impetus to the subject, but at the same time they have generated a remarkable amount of confusion and dissatisfaction among the experimentalists. The cause of this is perhaps the inability of the experimentalist to understand in a critical way the basis of the theoretical work. It is a difficult task for a non-expert to build up a critical picture of the situation concerning wave function computation as it affects the idea of aromaticity.

The semi-empirical approach to aromaticity has been well presented by Garratt in his excellent book and for this reason and because we favour the non-empirical approach to wave functions we have concentrated on the latter methods. This policy, together with our belief in separating theory from facts, has led to the development of a book with an unusual style. One advantage of this style is that as the theory grows it can be incorporated into this text without major changes being required as they are when a new semi-empirical approach is developed. Moreover, additional experimental facts can likewise be put alongside those already present without major changes in the organisation. It is against this background that we search for a definition of the word 'aromatic'.

We feel strongly on the point that one should in the first instance separate the facts from the theories. The danger in mixing the two together at the beginning is that it seems to be fatally easy to select those facts which do fit one's theories while neglecting those facts which do not. It is one of the major difficulties with the semi-empirical methods that this clear separation of fact and theory is not possible, because the parameters which enter the theory must first be got from the experimental facts and this process requires subjective judgements.

The general level of the book is perhaps best described as advanced undergraduate or research worker although we have naturally tried to make the book available to as wide an audience as possible.

One minor but irritating point is the question of the use of SI units. It is plain that the existence of a single unified set of units for science and technology is a desirable and useful thing. What is less clear is whether these units alone should be used all the time in all areas of all subjects, bearing in mind that SI units are undeniably more clumsy than are the natural units of a particular field. For example, in molecular structure theory, the natural units are defined by setting the electron charge as the unit of charge, the electron mass as the unit of mass and finally by setting Planck's constant divided by 2π as the unit of energy times time ($e=m=\hbar=1$). This system is so natural that it will surely persist indefinitely. Like other writers, we compromise here by using SI units where possible without serious distortion of common sense.

Finally, we hope that this book will generate in the experimentalist an enthusiasm for, and an interest in, the more formal and thorough methods of the quantum theory, because we believe that the future of the centrally important subject of theoretical chemistry lies in the main with these theories and methods.

May 1974

Contents

Chapter 1 GENERAL INTRODUCTION Introduction–historical background–general nature of aromaticity and preliminaries 1

Chapter 2 EXPERIMENTAL EVIDENCE ON AROMATICITY Introduction–molecular geometry–molecular energetics–spectroscopy–dipole moments–miscellaneous evidence–summary 10

Chapter 3 THEORETICAL IDEAS ON AROMATICITY Introduction–qualitative ideas–the ideal situation–the current situation–lower approximations–resonance energy–CNDO theory–σ-electrons as two-electron bonds –π-electrons-only theories–summary 37

Chapter 4 SIMPLE EXAMPLES OF AROMATICITY Introduction–phenanthrene–azulene–[18] annulene–ferrocene–thiophen–fulvene–pentalene–summary 58

Chapter 5 AROMATICITY OF SOME SELECTED CLASSES OF COMPOUNDS Seven-membered carbocycles–quinoid structures–4-pyrones–porphyrins and phthalocyanins–summary 71

Chapter 6 HOMOAROMATICITY 86

Chapter 7 ANTIAROMATICITY 90

Chapter 8 CONCLUSION AND COMMENTS 94

Appendix 96

References 97

Index 105

Acknowledgment

The authors are indebted to colleagues in the Department of Chemistry at Royal Holloway College for their guidance in the preparation of certain areas of this book.

1. General Introduction

Introduction – historical background of aromaticity – general nature of aromaticity – general Preliminaries

1.1 Introduction

Our purpose in writing this book at this time is to give a general account of the problem of aromaticity. There are many excellent discussions [1-18] on the subject so we confine ourselves to a *critical* examination of both experimental and theoretical aspects of the topic. We have tried to make the book self-contained and have also emphasised certain material which we consider is complementary to that normally available in the other reviews. If this leads to an unusual balance it must be realised that no one account can do justice to the entire subject of aromaticity. In particular, we have excluded consideration of inorganic aromatics. [19,20]

In dividing up the material we have kept in mind that it is important to distinguish experimental fact from theoretical conclusions in the first instance. The confrontation of the two should form a second and distinct stage. Accordingly, after a brief historical introduction in this chapter, we have two major chapters, the first on experimental facts (chapter 2) and the second on theoretical ideas (chapter 3). In the following two chapters, we compare theory and experiment using first some simple clear examples (chapter 4) and then some more involved examples in which whole classes of molecules are discussed (chapter 5). Homoaromaticity (chapter 6), and anti-aromaticity (chapter 7) are considered and there is a summarising chapter (chapter 8).

We have examined the literature generally up to the end of 1972 and some later material has also been included.

1.2 Historical Background of Aromaticity

There are several good historical accounts [4,9,16] of the concept of aromaticity and the following remarks are intended as a brief summary.

The general idea of aromaticity arose when it was noticed in the nineteenth century that carbon compounds can be put into two groups, aromatic or aliphatic, *that is*, non-aromatic. The two groups of compounds differ in their physical and chemical properties to a sufficient extent to make the distinction a useful one [21, 22]. One particular difficulty with the aromatic compounds was that they analysed as unsaturated compounds and yet they were chemically different from alkenes or alkynes.

The first real understanding of this situation developed as the general ideas about the two-electron bond were formulated [23] in the 1910s and 1920s. It became clear at that time that the aliphatic compounds have an electron organisation which is well described by the two-electron chemical bond. It then follows that the electron organisation of the aromatic compounds must differ in some important way from that of the aliphatic compounds. The suggestion of the aromatic sextet [24] led E. Hückel in 1931 to the fundamental idea of the π-electron molecular orbitals and the famous $(4n+2)$ rule of aromaticity.[25] The competing theory of the 1930s was the idea of resonance [26] between valence bond structures as in figure 1. In recent years, this theory has proved less useful than the molecular

la lb

Figure 1.

orbital theory, partly because it is difficult to make detailed computations for large molecules where there are many contributing structures and partly because it is difficult to interpret the resulting wave functions when you have them. We accordingly use the molecular orbital theory throughout. There is however a complication here in that many valence bond ideas have become part of the language of chemistry and have become mixed with molecular orbital theory. One such is the concept of cross conjugation[26a] as in (molecule 2a) as compared with simple conjugation as in (molecule 2b). This has no clear counterpart in molecular orbital theory. A case in point is that of fulvene (molecule 2c) which is frequently called cross conjugated.

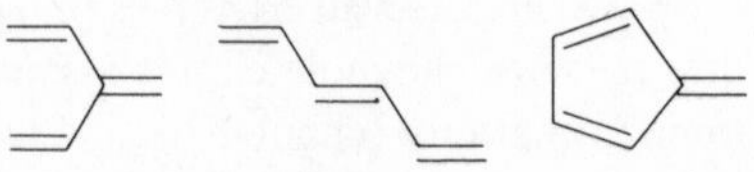

2a 2b 2c

Figure 2.

The 1940s and 1950s saw the extensive development of the simple π-electron molecular orbital theory and much of this was concerned with aromatic systems.[27,28] Up to the end of the fifties, aromaticity was still largely confined to benzene and its homologues, but developments in molecular orbital theory then sparked off much interesting non-benzenoid and small ring carbon chemistry such as the examples shown in

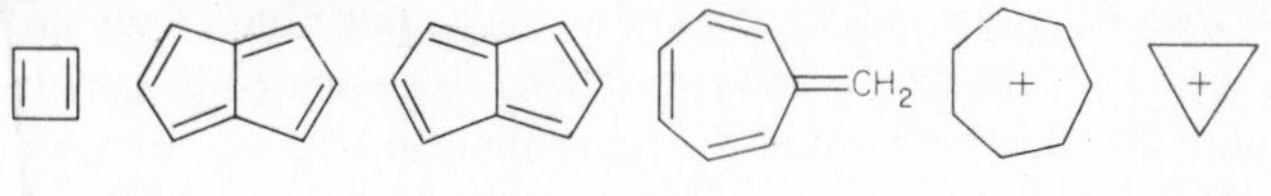

3 4a 4b 5 6 7

Figure 3.

figure 3. This is one of the really effective impacts which theory has made on chemistry in the last decades.

The 1960s are notable for the general spread of molecular orbital theory throughout chemistry and the extensive development of the more elaborate, but not yet really accurate, methods for dealing with π-electron systems.[29]

Although the idea of aromaticity has existed for some years, some fundamental questions remain open. For example, should we be able to divide all known carbon compounds into two groups, [30] one aromatic and one non-aromatic? To do this needs a theoretical or experimental criterion of aromaticity which is generally accepted. No such criterion is available now.

1.3 General Nature of Aromaticity

In this section, we try to give an interpretation of the term 'aromatic'. An ideal definition would be one based on experimental fact and not on theory. The advantage of such a definition is that it does not change with time as theories change and one is spared the spectacle of molecules being reclassified every few years.

One such definition is that suggested by Albert[31] and based on bond lengths. To express Albert's criterion loosely, we say that a ring is aromatic if its carbon-to-carbon bond lengths are like those of benzene (*cf* chapter 2.2).

But there are difficulties with this definition of aromaticity. One is that a particular bond or bonds may be long (as is the transannular bond of azulene) in a molecule which is widely regarded as aromatic. Another difficulty is the extension of the definition to hetero-atom systems. But perhaps the greatest difficulty is that one simply does not know the bond lengths in the vast majority of cases.

One classical definition of aromaticity which has a high experimental content is the thermodynamic one that an aromatic molecule is more stable, *that is,* of lower total energy, than one would expect if it were composed of simple two-electron chemical bonds. Until recently, we have had no clear understanding of the term 'simple two-electron bonds' but there seems to be some hope in this direction now. [33]

It is certainly true, however, that most of the definitions of aromaticity of the 1910 to 1920 period were in terms of this thermodynamic criterion [34] which later grew into the idea of 'resonance energy' (*cf* chapter 2.3 (i) and 3.6).

Another definition of aromaticity which is largely experimental is that based on the magnetic properties of certain cyclic and aromatic looking molecules. [35] For example, the presence of a ring current in a molecule is associated with magnetic anisotropy and with the positions of the proton signals in the n.m.r. spectrum. The ring current is generated by simple circulation of the electrons round the ring (diamagnetism) or by distortion of the electron clouds (paramagnetism) so that the proton signal appears in a different region from that expected in the absence of a ring current. Equally, diamagnetic anisotropy is enhanced in one direction by these currents and this is either observed by measuring the three principal components of the diamagnetic tensor or, indirectly and more easily, by its effect on the average of the three components of the diamagnetic tensor. In addition, coupling constants (*J*-values), are often held to reflect the degree of aromaticity (*cf* chaper 2.4).

The difficulty with the n.m.r. criterion is that one cannot be sure where the proton signal would be were the ring current not present and, although this definition of aromaticity looked most useful in the early stages,[36] and·is still used, some objections have been raised to its use. [37, 38] In particular, a study[39] of how complex the interpretation of n.m.r. spectra can be suggests that care is required in using these magnetic criteria of aromaticity.

Another experimental definition of aromaticity is that the compound 'has a chemistry like that of benzene'. In other words, undergoing substitution rather than addition reaction. Thus with molecular bromine, benzene gives the substitution product (**9**) rather than the addition product.

8 + Br_2 —Catalyst→ 9 (Br)

Figure 4.

This definition is simple, readily applied and seems to fit the general feelings of many 'experimental' chemists. [40, 41] One difficulty with it is that it depends on reaction conditions, *for example,* catalysts. A second difficulty is to link this definition with theoretical considerations, although attempts have been made to do this. [9, 41]

Other experimental criteria for aromaticity have been suggested [42] (*cf* chapter 2) but no one of these seems to have reached general acceptance at the present time.

In trying to summarise the situation, one point at least is clear – the special place of benzene. In so far as there is any agreement as to what aromaticity means it seems to lie in the phrase 'like benzene' and we propose to use this as our basic definition of aromaticity. This by no means solves all our problems because this term is merely a label which classifies a molecule as a member of a particular set of molecules and so leaves the definition empty of real meaning. This definition has no theoretical content so that future changes in our understanding of the benzene wave function will not affect it. The force of this argument may be seen if it is remembered that in the 1930s aromaticity was defined[40] in terms of resonance energy between valence bond structures, as in figure 1, whereas now it is commonly defined in terms of π-electron molecular orbitals. Perhaps in ten years time it will be defined in terms of the thirty σ- plus π-electrons of benzene, and some of the results of the accurate, all electron calculations (chapter 3.4) support this suggestion. But the use of benzene as a reference standard looks inescapable for the foreseeable future.

At this point we begin to introduce some theoretical concepts. At the lowest level, we say that a molecule or ion shall be called aromatic if its electron organisation is like that of benzene without being explicit about what that of benzene really is.

There is also the opposite and relatively neglected question of the definition of the term 'non-aromatic' or 'aliphatic'. It has often been held to be obvious that such molecules are those whose electron organisation can be described accurately by two-electron chemical bonds. Thus, the saturated and unsaturated molecules such as methane (**10**) or

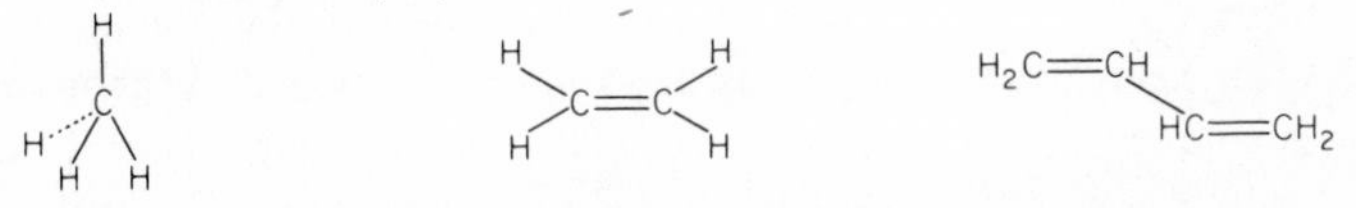

$H_2C{=}CH{-}HC{=}CH_2$

10 11 12

Figure 5.

ethylene (**11**), butadiene (**12**) *etc* are not aromatic because the electron organisation of each is reasonably accurately described by a single structure as shown in figure 5. But if we are to discuss aromatic molecules in terms of molecular orbital theory, surely we should also discuss the non-aromatic ones in the same language. This has been difficult until recently but it is now possible to compute the localised bonds *quantitatively* and rigorously[33] and to describe the electron organisation in simple two-electron bonds to any desired accuracy. When this work has been developed, a detailed understanding of the 'non-aromatic' molecule will be available and this should throw light on the counter problem of the bonding in the aromatic molecules. We feel that much of the dispute over aromaticity is as much a dispute over the non-aromatic reference molecule as over the aromatic molecule.

For the moment, we describe as non-aromatic a molecule whose electron organisation is accurately described by localised two-electron bonds. However, we cannot make the contrary statement that a molecule is aromatic because its electron organisation is described accurately by non-localised molecular orbitals. Such a definition would include many systems which are not commonly thought of as being aromatic, *for example,* excited states of a molecule, transition state of a reaction, delocalised bonding as in the nitro group. Evidently we need to consider a special type of delocalised bonding.

If we continue with our definition of an aromatic molecule as one which has an electron organisation like that of benzene, one immediate corollary is that any aromatic system will be cyclic. It hardly seems possible to imagine an electron organisation like that of benzene in an acyclic system and it is generally accepted that aromatic systems are cyclic systems.[1-18] Even this point might be challenged by the consideration of a transition state which is half way between cyclic and acyclic and which seems to be aromatic in some sense,[43] but this is a special case which arises the entire question of the nature of transition states and we will not continue this point further here.

It is necessary to consider two further aspects. The first is that aromaticity is due only to the π-electrons. This seems today to be too restrictive, not only because we now understand that the σ-electrons play a major role in determining the π-electron organisation, as was established by the all-electron computations on some small molecules,[44] but also because it is far from certain that it is the π-electrons rather than the σ-electrons which confer the aromaticity on benzene. Moreover, in the inorganic aromatics the *d*-orbitals seem to play a part, and in situations such as that of homoaromaticity the electrons are not clearly either σ or π in nature. For these reasons we suggest that the concept of aromaticity cannot now be restricted to π-electrons only.

The second point is the planarity or near planarity of the ring atoms. This again is often regarded as a requirement for aromaticity and as far as we know there exists no clear examples of aromaticity in systems which are far from planar. There is a complication here in that it is sometimes difficult to distinguish the geometry of the molecule from that of the ring system in such cases as bicyclo-aromaticity[45] or spiroconugation.[45a]

This brings us to the famous Hückel definition of aromaticity, the $(4n+2)$ rule.[25] This rule is simply that rings of $p\pi$ atomic orbitals with 6, 10, 14 .. electrons in a closed shell ground state are considerably lower in energy than are the rings with 4, 8, 12 ... electrons. This rule has been used with the case of n equal to zero in the cyclopropenyl cation. The result is implied by the energy level diagrams as shown later in figure 31. Strictly speaking, the rule only applies to planar monocycles but it is often used in a wider context.

The rule has been modified by later workers who suggest that the stability of the $(4n+2)$ cycles falls off as n increases until *ca* n=5 the cyclic polyene with alternating single and double bonds is the stable form, (chapter 3.8).

Other theoretical definitions of aromaticity are mentioned in the theoretical section. One definition will not be discussed there and merits special mention here. This is the purely theoretical definition due to Craig.[46] In formulating the valence bond wave function of both benzenoid and non-benzenoid hydrocarbons, he noticed that some of the latter seem to have a ground state wave function which does not transform as the fully symmetrical (all+1 or identity) irreducible representation of the symmetry group of the molecule. Thus, in the case of pentalene in figure 3, if we assume D_{2h} symmetry of the nuclear framework, then the wave functions of the two Kekulé structures have the symmetry B_{1g} rather than A_g of D_{2h}. This cannot happen with molecular orbital wave functions because, if the electrons form closed shells so that no molecular orbital is singly occupied, then the many-electron wave function is, of necessity, fully symmetrical.

It is far from clear just what this result means. Perhaps the nuclei do not adopt D_{2h} symmetry.[47] Perhaps the Kekulé structures are only minor contributors to the ground state wave function which could be a mixture of ionic and long-bonded structures. Or perhaps the molecular orbital point of view is quite wrong here and the filling of the molecular orbitals to form the lowest energy state is incorrect. This last possibility bears investigation because our entire molecular orbital theory is based on the validity of this filling of the lowest energy molecular orbitals first. Only if the total energy were the sum of one-electron energies, which it is not (see chapter 3.5 *etc*), would this procedure be exact.

Craig's definition does suffer from the restriction of applying only to symmetrical molecules and also that it is not easy to connect it with measurable properties of the molecule. Ambiguities may arise in its use but it has certainly led to much interesting further work.

A possible test of how well one understands the term 'aromatic' is to take the available criteria of aromaticity and apply them in an area other than those areas in which they were worked out originally and where they obviously perform well. One such new area is the electronic excited states of, say, benzene and the question is whether we should think of these as being aromatic in some sense.

We suppose as usual [48a] that the excited states of benzene are generated by π-electron excitations so we have diagrams such as that of figure 6 to represent the configurations of the excited states.

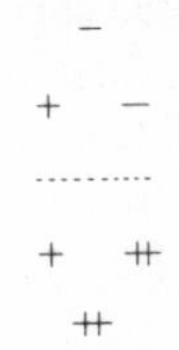

–

\+ –

\+ ++

++

Figure 6. π-Electron energy levels for excited state benzene (vertical axis: orbital energy)

This π-electron arrangement plainly differs drastically from that of the ground state, so if the π-electrons determine the aromaticity then the excited states can hardly be aromatic species. But if the σ-electrons determine the aromaticity and if the rearrangement of the π-electrons does not affect the σ-electrons appreciably, then the excited states will be aromatic. It is possible that either the excited states or the aromaticity question involves a mixture of σ and π-electrons and thus there is no clear conclusion to be drawn in this way.

A closely related test is that of the geometry of the molecules in the excited states. It is well established [48b] that the $^1B_{2u}$ state (the upper state of the 2600 Å band) is a flat regular hexagon (D_{6h}) but the lowest triplet state which is probably $^3B_{1u}$ may or may not be distorted from the D_{6h} geometry.[48c] The bond lengths of the $^1B_{2u}$ state are 1.435 Å for the carbon-carbon bond and 1.07 Å for the carbon-hydrogen bond (the ground state values are 1.397 Å and 1.084 Å respectively [48b]).If we suppose, with Albert, that the equality of bond lengths is a guide to aromaticity then the $^1B_{2u}$ could be described as aromatic. And even the ${}^3B_{1u}$ state could be called aromatic if it is a regular hexagon and so like benzene in that sense.

The chemical reactivity criterion is difficult to apply unless one simply says that the reactivity of the excited states is very high so the molecule is not aromatic. The magnetic tests would be helpful here but they are experimentally difficult to apply. The thermodynamic test seems to be inapplicable because there is no clear comparison that one can make in an effort to see whether the energy of the molecule is 'lower than expected'. The azulene molecule is of interest in this respect (*cf* chapter 4.3).

There are two other areas which might be described as new and so be a useful testing ground for our theories. One [49] is the benzyne molecule. By inspection, this molecule closely resembles benzene and it might seem that it will have an electron organisation close to that of benzene. It is thought both on experimental and theoretical grounds that there may be considerable resonance energy in the molecule although the considerable strain energy which the molecule must have makes it difficult to be sure about this. From these results, the molecule might well be aromatic. On the other hand, the chemical reactivity of benzyne is extremely high, the magnetic criteria are inapplicable and the geometrical ideas meet with difficulties in the matter of the short triple bond. So the overall picture is quite unclear.

The other case is the porphin molecule. This will be discussed in detail later (chapter 5) but the point which concerns us now is that this molecule has been found (table 12) to have a resonance energy of 504 kcal mol^{-1}. The experimental method used was the well established combustion method which gives the heat of formation (chapter 2). If we say that four times the resonance energy for pyrrole cannot be greater than about 150 kcal mol^{-1}, then the bridges seem to be contributing a very large amount of resonance energy indeed. It is of course possible to develop calculations which contain disposable parameters and which reproduce such resonance energies but their value seems doubtful to us.

It seems that if the idea of trying out criteria of aromaticity in a new area is a realistic test of the present state of affairs, then we do not know very much about the central question of 'aromaticity'.

1.4 General Preliminaries

One major difficulty with this work has been the question of how to represent the π-electron organisation in aromatic molecules. There has been much confusion generally over this point, but the mainstream practice seems to be to use a Kekulé structure to represent the real molecule. We have generally followed this practice in all neutral molecules save benzene itself. For the latter, the real molecule is drawn with an inscribed circle, except in figure 1 where the two Kekulé structures are drawn. Elsewhere, a Kekulé structure for benzene drawn as a regular hexagon represents the hypothetical molecule which is cyclohexatriene with all six carbon-to-carbon bonds of equal length. A cyclic ion is represented with the polygon with the sign of the charge inscribed within. It is worth mentioning that there exists no formal solution to the basic difficulty because it is plainly impossible to represent the electron organisation in a molecule with a simple picture of this kind.

2. Experimental Evidence on Aromaticity

Introduction – molecular geometry – molecular energetics: ground state energetics, equilibria energetics, chemical reactivity energetics – spectroscopic: nuclear magnetic resonance (n.m.r.), magnetic susceptibility and anistropy, electron spin resonance (e.s.r.), vibration, electronic, photoelectron (P.E.S.) – dipole moments – miscellaneous – summary and comments.

2.1 Introduction

All of the properties of a molecule are potential sources of information about its aromaticity. To say the same thing in theoretical terms, the wave function of a molecule determines all the properties of a molecule and, conversely, all the properties of a molecule throw some light on its wave function and hence its aromaticity. No doubt some properties will be more sensitive to the aromatic nature of the electron organisation than will others.

In this chapter, we survey briefly some of the properties of a molecule which seem likely to have a direct and clear bearing on whether or not it is aromatic.

2.2 Molecular Geometry

The geometry of a molecule can be direct and convincing evidence that it is like benzene and therefore aromatic. One aspect of a molecule's geometry is its bond lengths and Albert[31] suggested that '... of two molecules, the more aromatic will have the longer double bonds and the greater equalisation of carbon-carbon bond lengths throughout the molecule'. Some refinements of this criterion have been proposed.[50, 51]

One refinement[50] is that the molecule is aromatic if its carbon-carbon bonds are between 1.36 and 1.43 Å in length, while the molecule is a polyene if it has alternating bond lengths of 1.34 to 1.356 Å for the double bonds and 1.44 to 1.475 Å for the single bonds. To illustrate this criterion of aromaticity, the carbon-carbon bond lengths[52] of the fulvene (**13**) are compared with those[53] of the non-aromatic molecule butadiene (**12**) and with those[54] of cyclopentadiene (**14**) in figure 7. The results show that the fulvene is a polyene.

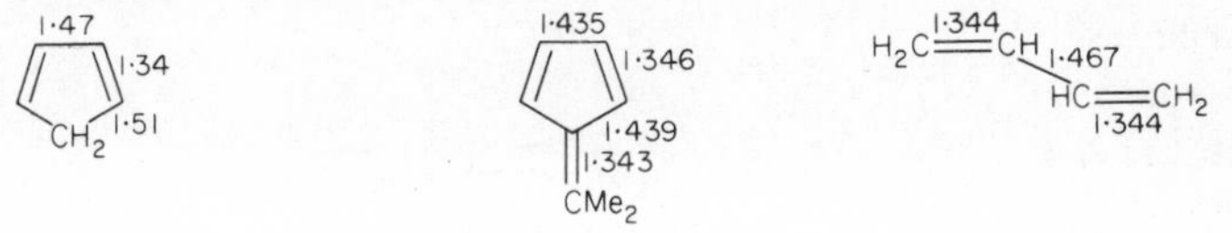

1·47 1·34 CH_2 1·51

1·435 1·346 1·439 1·343 CMe_2

H_2C=CH 1·344 1·467 HC=CH_2 1·344

14 13 12

Figure 7.

As a contrast, the bond length[55] of benzene (**8**) is compared with those of naphthalene (**15**) in figure 8. It is obvious that naphthalene is indeed an aromatic molecule on the bond length criterion. Further examples of this criterion are discussed in chapters 4 & 5.

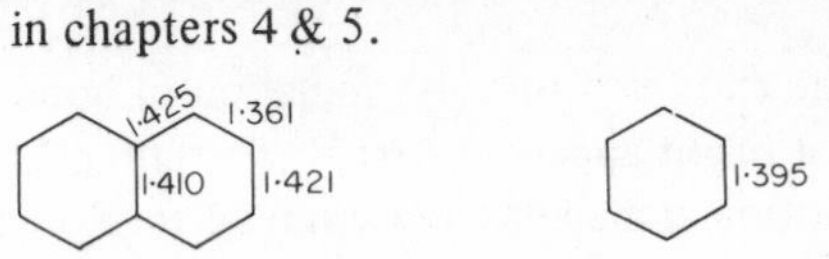

1·425 1·361 1·410 1·421

1·395

15 8

Figure 8. Bond lengths (Å) for naphthalene[a] and benzene[b]

a. ref. 55b, M154s

b. ref. 55

The second refinement[51] of Albert's criterion is a more quantitative statement based on taking the mean square deviations of the carbon-carbon bond lengths as a measure of aromaticity. The mean bond length of the n bonds, $\bar{d}$, is defined by

$$\bar{d} = n^{-1} \sum_{(rs)} d_{(rs)}$$

where $d_{(rs)}$ is the length of the rs bond. The mean square deviation σ is defined as the dimensionless quantity

$$\delta = n^{-1} \sum_{(rs)} (d_{(rs)} - \bar{d})^2 / \bar{d}^2$$

Then define the aromaticity constant, A, by

$$A = 1 - 225\delta$$

Some typical results are given in table 1 and generally these seem consistent with other criteria of aromaticity.

Table 1. *Aromaticity constants based on bond lengths*[a]

Molecule	Aromaticity Constant (A)
benzene	1.00
naphthalene	0.90
phenanthrene	0.93
anthracene	0.89
azulene	1.00
thiophen	0.93
pyrrole	0.91
furan	0.87
fulvene	0.62
butadiene	0.62

(a) Taken from ref. 51 except last entry.

It should be made clear that the determination of bond lengths is by no means so simple as it might appear at first.[56] The complication of the zero-point vibrational motion and energy is a serious one. Roughly speaking, the zero-point vibrational energy expressed in kcal mol^{-1} is equal to the number of bonds in the molecule if one supposes that the average vibrational quantum is about 3 kcal mol^{-1} (about 1000 cm^{-1}).

To see the order of magnitude of the problem created by the zero-point vibrational energy, consider a benzene carbon-carbon bond which during a complete vibrational motion will change in length by about 0.1 Å. This number is twice the amplitude for a classical oscillator with the physical characteristics of this bond. The situation is less serious with a quantum oscillator because the probability distribution for the bond length has a maximum at the equilibrium bond length, and not at the extremes of bond length as does the classical oscillator.[57]

Unless the vibrational behaviour is allowed for,[58] a measured bond length may only be accurate to $\pm$0.01 Å. It follows that differences of about 0.01 Å ought not to be used as a basis for arguments. We stress this difficulty of the significance of small differences in bond lengths because exaggerated claims are sometimes made for the accuracy and significance of some bond length meausrements.

Another geometrical feature of a molecule which is relevant to the question of aromaticity is the planarity or near planarity of the ring atoms. It is less easy to make quantitative statements on this topic than about bond lengths but there is certainly a tendency for aromaticity to be accompanied by planarity. A good example of this point is provided by the large annulenes[59, 60] where the aromatic [*18*] annulene is nearly planar while the non-aromatic [*16*] annulene is significantly non-planar.

In the light of this discussion about molecular geometry, it is clearly desirable to obtain geometrical information in a routine manner and this is now being done, but, as far as we know, there is no information available on internuclear distances for any hydrocarbon anion (figure 9) and, of the cations, only the C_3 and the C_7 cycles seem to have

16 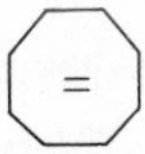 17

Figure 9.

been measured. This particular examples are triphenylcyclopropenyl perchlorate (**18**)[61] and preliminary information on the tropylium cation

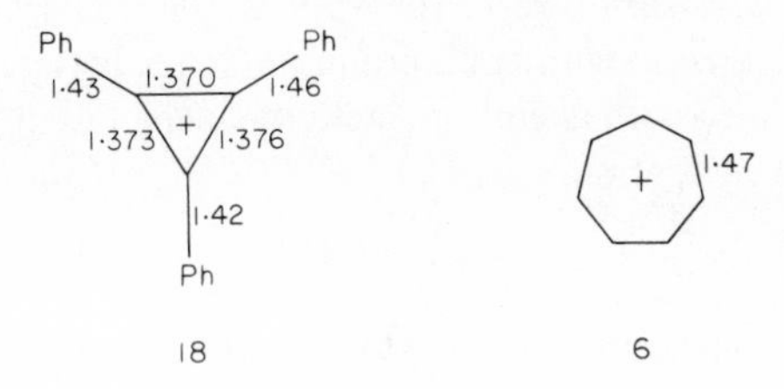

Figure 10.

itself (**6**)[62] as shown in figure 10. There are major experimental difficulties in both measurements and particularly in the tropylium ion case, due to rotation of the cation in the crystal, so that the result should be accepted only with serious reservations. Indeed it seems to us that the tropylium value is too large.

The result for the substituted cyclopropenyl cation (**18**) is striking in that the carbon-carbon bond length is nearly the same as that in benzene but the result could be fortuitous. Indeed, if the result is not fortuitous and the electron arrangement is approximately the same in benzene and in the cyclopropenyl cation to give the same bond length, we will have to revise our ideas about the σ-bonding in such molecules.[63]

To summarise, we feel that aromaticity defined in terms of bond lengths is a more or less valid criterion. Unfortunately we rarely have accurate geometrical information about the molecule we are discussing. It is interesting to reflect on the huge number of organic molecules, about which we have all kinds of sophisticated kinetic information, while the geometry of the molecules is, to some extent at least, a matter of guesswork. So we reluctantly discard molecular geometry as a general and practical criterion of aromaticity although it should be used whenever it is possible to do so.

A possible way around the difficulty of not knowing the actual bond lengths is to use the bond order concept[64] and to say[65] that a molecule is aromatic if all its Hückel bond orders are between 0.4 and 0.8. More sophisticated theoretical methods[66] may be used but, since the bond order idea is itself a crude one, perhaps the Hückel bond orders are as useful as any.

Another possibility is that bond lengths can sometimes be inferred from the values of coupling constants from the n.m.r. (chapter 2.4).

One area in which the geometrical criterion of aromaticity may prove particularly useful in the future is the question of the aromaticity of excited states. Other criteria are very difficult to apply in this region, either for experimental or for theoretical reasons.

One interesting theoretical point which should be mentioned is that the first signs of electron delocalisation is a shortening of the single bond or bonds with no lengthening of the double bond or bonds.[67]

2.3 Molecular Energetics

We subdivide this material into three subsections, the first dealing with ground state energetics, the second with equilibrium processes between two stable molecules and the third with chemical•reactivity energetics in terms of ground and transition state energetics.

2.3.1 Ground State Energetics (thermochemical data)

Energetic information about the ground state of a molecule is usually recorded in one of two ways. A thermodynamicist may report the standard heat of formation of a compound, ΔH_f° (temp), which is the energy required to form the molecule in its standard state from the elements in their standard states. Thus, the standard heat of formation of liquid benzene, ΔH_f° **(298)**, is the energy required to form one mole of liquid benzene at 298 K from solid carbon (graphite) and gaseous hydrogen at 298 K and unit pressure. This number is $\Delta H_f^\circ(298) = 19.82$ kcal mol^{-1} and is positive by convention[68] since the formation reaction is endothermic:

$$6\text{C(graphite)} + 3\text{H}_2\,\text{(gas)} \longrightarrow \text{C}_6\text{H}_6\,\text{(liquid)};\ \Delta H_f^\circ = 19.82 \text{ kcal mol}^{-1}$$

This thermodynamicist's standard heat of formation is less useful for theoretical purposes than is the 'atomisation energy' or 'heat of atomisation', ΔH_a (temp), of the molecule. The latter quantity differs from the standard heat of formation only in that it uses as reference point the gaseous atoms and not the elements in their standard states as does ΔH_f°. Thus, in hydrocarbons, the reference point for the standard heat of formation is graphite and gaseous hydrogen molecules while the reference point for the atomisation energy is the isolated gaseous carbon atoms and hydrogen atoms. The equation which connects the two quantities is, in the particular case of benzene

$$\Delta H_a^{\circ}\,(298) = 6 \times \Delta H_f^{\circ}\,(\mathrm{C,g})\,(298) + 6 \times \Delta H_f^{\circ}\,(H,g)\,(298 - \Delta H_f^{\circ}\,(298)$$

$$= 6 \times 170.9 + 6 \times 52.09 - 19.82$$

$$= 1318.1 \text{ kcal mol}^{-1}$$

where ΔH_f° (298) is the standard heat of formation at 298 K of liquid benzene. Heats of atomisation are always positive in agreement with the thermodynamics sign convention.

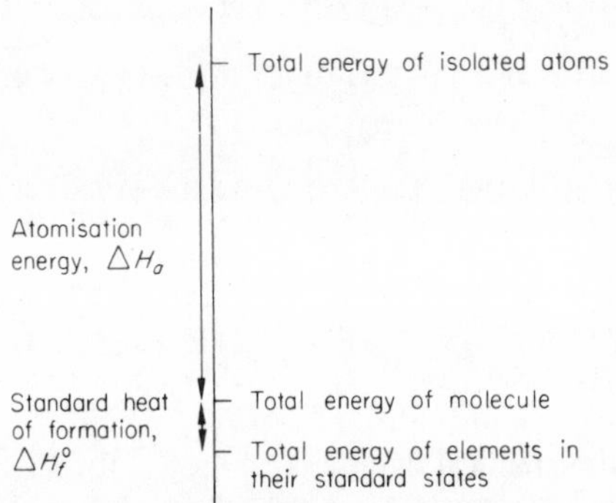

Figure 11. The diagram refers to a compound with a positive H_f°

The main advantage of expressing the result in terms of atomisation energy is that the resulting number is approximately the sum of the bond energies. Thus, for the twelve bonds of benzene with bond energies of some 100 kcal mol^{-1}, the atomisation energy is about 1200 kcal mol^{-1}. The number 19.82 kcal mol^{-1}, on the other hand, has no significance beyond being the heat of formation. A diagram showing the connection between heats of formation and atomisation energies is given in figure 11. The standard heat of formation may be positive or negative but is nearly always small. Some typical atomisation energies for hydrocarbons, together with the energy quantities from which they are obtained, are given in table 2. These results are in the main those subsequent to Wheland's compilation.[12] Extensive compilations[68a, 68b] are available.

The classical experimental methods[69,70] of establishing the ΔH_f° of a hydrocarbon are combustion to carbon dioxide and water and, for an olefinic hydrocarbon, hydrogenation to the saturated hydrocarbon. In the combustion experiment, the only unknown quantity is the total energy of the organic molecule and so this is obtained directly from the thermochemical equation.

In the hydrogenation experiment, there are two unknowns, the total energy of the unsaturated molecule and that of the saturated molecule. So one must know the total energy of the saturated molecule before the total energy of the unsaturated molecule becomes available. Sometimes this quantity is known experimentally and sometimes it is estimated by assuming additivity of bond energies plus some correction terms.

Table 2. *Standard heats of combustion, formation, sublimation and atomisation at 298 K*

Hydrocarbon	$-\Delta H^o_{comb}$(c)	ΔH^o_f (c)	ΔH^o_{sub}	ΔH^o_f (g)	ΔH^o_a (g)
acenaphthene (i)	1487.0[a]	+ 16.8	20.6[a]	+ 37.4	2534.3
acenaphthylene (ii)	1446.5[a]	+ 44.6	17.4[b]	+ 62.0	2405.5
anthracene (iii)	1689.17[c]	+ 30.87	24.3[c]	+ 55.2	2858.3
[18] annulene (iv)	2346.8[d]	+ 39.0	–	–	–
azulene (v)	1265.1[e]	+ 51.3	18.4[b]	+ 69.7	2056.0
bibenzyl (vi)	1807.24[c]	+ 12.30	20.1[f]	+ 32.4	3089.5
biphenyl (vii)	1494.22[c]	+ 24.03	19.5[b]	+ 43.5	2528.2
biphenylene (viii)	1486.3[h]	+ 84.4	20.0[b]	+104.4	2363.1
cyclopropene (ix)	485.0(g)[i]	–	–	+ 66.6	654.8
6,6-dimethylfulvene (x)	1116.1(1)[j]	+ 22.11 (1)	10.6(vap)[j]	+ 32.7	1855.4
6,6-diphenylfulvene	2243.7[j]	+ 72.6	–	–	–
fluoranthene (xi)	1891.77[k]	+ 45.37	23.7[b]	+ 69.1	3186.2
indene (xii)	1146.16 (1)[l]	+ 26.44 (1)	12.64 (vap)[l]	+ 39.1	1838.9
naphthalene (xiii)	1232.35[c]	+ 18.57	17.42[m]	+ 36.0	2089.7
perylene (xiv)	2334.60[k]	+ 43.69	30.0[n]	+ 73.7	3969.4
phenanthrene (xv)	1686.06[c]	+ 27.76	21.7[b]	+ 49.5	2864.0
pyrene (xvi)	1873.83[k]	+ 27.44	22.5[g]	+ 49.9	3205.4
triphenylene (xvii)	2136.53[k]	+ 33.72	28.2[p]	+ 61.9	3639.4

The table gives thermochemical data, in kcal mol^{-1}, for a selection of cyclic hydrocarbons. Most of the data is later than that given by Wheland (ref. 12). The roman figures in column 1 refer to the diagrams. Columns 2 and 3 refer to the crystalline state unless otherwise specified. (1) = liquid and (*g*) = gas. In column 4 (vap) is the heat of vapourisation while ΔH^o_{comb}, ΔH^o_f, ΔH^o_{sub} and ΔH^o_a are the standard heats (at 298°) of combustion, formation, sublimation and atomisation respectively.

References for Table 2

a. Boyd, R.H., Christensen, R.L. and Pua, R. (1965) *J. Amer. Chem. Soc.* **87,** 3554

b. Morawetz,E. (1972), *J. Chem. Thermodynamics* **4,** 455

c. Coleman, D.J. and Pilcher, G. (1966) *Trans. Faraday Soc.* **62,** 821

d. Beezer, A.E., Mortimer, C.T., Springall, H.D., Sondheimer, F. and Wolovsky, R. (1965) *J. Chem. Soc.* 216

e. From the data of: Quitzsch, K., Schaffernicht, H. and Geiseler, G., (1963), *Zeit. Physik Chem (Leipzig)* **223,** 200.

f. Aihara, A. (1959) *Bull, Chem. Soc. Japan,* **32,** *1242*

g. Bradley, R.S. and Cleasby, T.G. (1953) *J. Chem. Soc.* 1690

h. Bedford, A.F., Carey, J.G., Millar, I.T., Mortimer, C.T. and Springall, H.D. (1962) *J. Chem. Soc.,* 3895

i. Wiberg, K.B., Bartley, W.J. and Lossing, F.P. (1962) *J. Amer. Chem. Soc.* **84,** 3980

j. Day, J.H. and Oestreich, C. (1957) *J. Org.Chem.* **22,** 214

k. Westrum, Jr., E.F. and Wong, S. (1967) *Thermodynamik-Symposium,* Paper II-10, Heidelberg; U.S. At. Energy Comm. COO- 1149 – 92; (1967) *Nucl. Sci. Abstr.* **21,** (19), 34248; (1968) *Chem. Abstr.,* **68,** 95236

l. Stull, D.R., Sinke, G.C., McDonald, R.A., Hatton, W.E. and Hildenbrand, D.L., (1961) *Pure Applied Chem.,* **2,** 315. [*cf* book by Cox and Pilcher ref.68a].

m. Irving, R.J. (1972) *J. Chem. Thermodynamics* **4,** 793

n. Wakayama, N. and Inokuchi, H. (1967) *Bull. Chem. Soc. Japan,* **40,** 2267

p. Hoyer, H. and Peperle, W. (1958) *Z.Elecktrochem.* **62,** *61.*

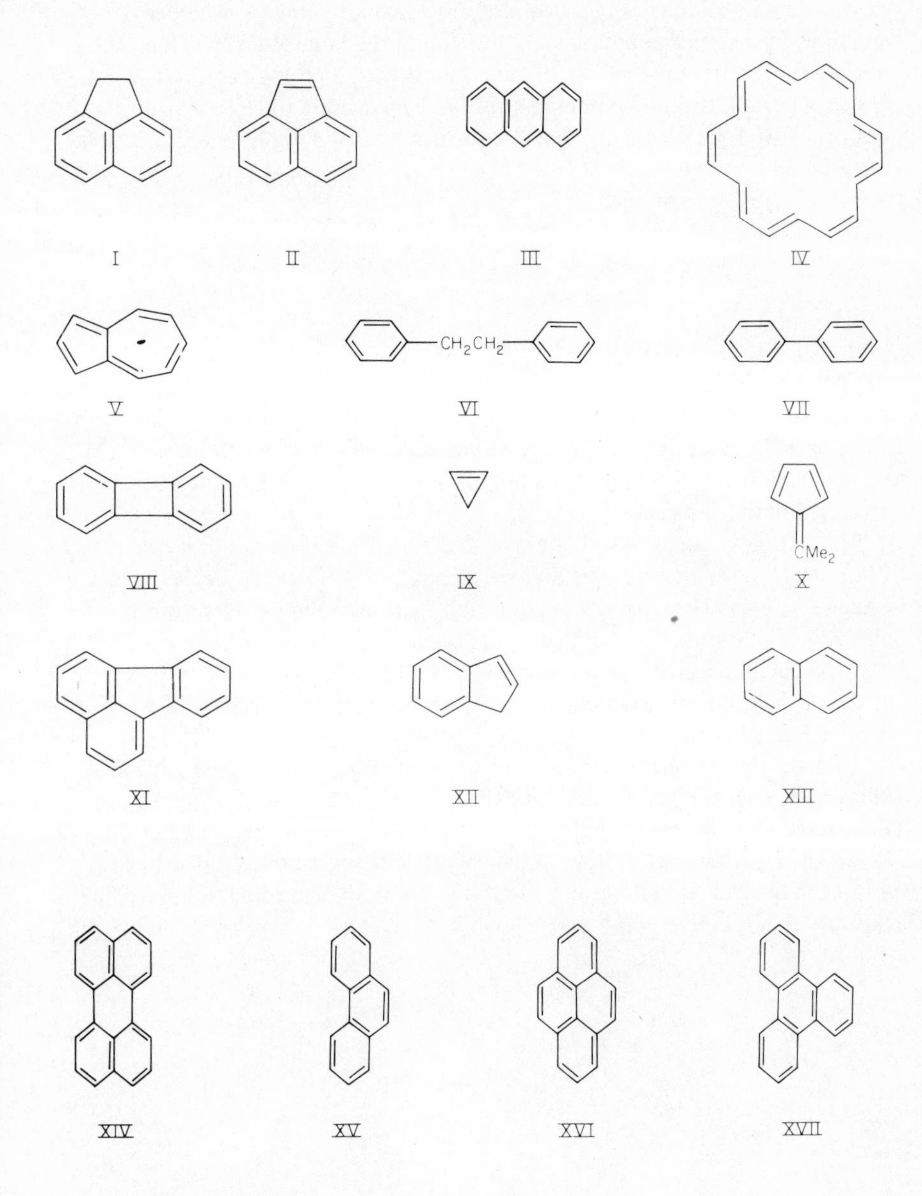
I
II
III
IV
V
CH_2CH_2
VI
VII
VIII
IX
CMe_2
X
XI
XII
XIII
XIV
XV
XVI
XVII

Once we know the total energy of a number of molecules, we can compare them to see whether a given molecule has a total energy which we would expect were its electrons organised into simple bonds – that is, whether the atomisation energy is approximately the sum of the bond energies. Thus, we might compare the total energy of benzene (**8**) with that of cyclohexatriene (**19**) (figure 12). If the two numbers are the same, within the experimental errors involved, we would say that the electron organisation of benzene is well represented by the formula (**19**).

1·395 1·395 8 1·395 1·395 19

Figure 12.

If, as is the case, the total energy of real benzene is lower than that of (**19**). we conclude that (**19**) is a poor representation of the way the electrons are organised in this molecule.

The difficulty which has arisen here, and which has given rise to much controversy in the literature, is that we do not have available to us the molecule cyclohexatriene (**19**). This is a hypothetical molecule, and we must somehow estimate what its energy would be if it existed.

A new and interesting possible way out of this difficulty is via complexes[71] in which the three double-bonds are localised but further analysis of this new approach is needed.

To make up cyclohexatriene, we might try taking three ethylene molecules, breaking the appropriate six carbon-hydrogen bonds and then assembling the three fragments into the cyclohexatriene molecule. The process is depicted in figure 13. Then we must follow all the energetic changes in this process and this is the difficult step because there is no universally agreed method of doing this. We do not know, with

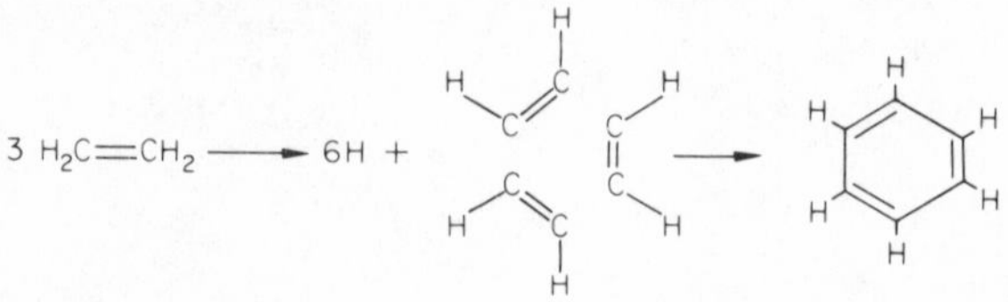

$3\ H_2C{=}CH_2 \longrightarrow 6H +$ [three HC=CH fragments] $\longrightarrow$ [cyclohexatriene, C_6H_6]

11 20

Figure 13.

certainty and accuracy, what numbers to write down for the individual bond energies and for the various corrections which must surely arise. No doubt there will be changes in bond lengths, in hybridisations, in non-bonded repulsions, in electrostatic terms, in the sigmaconjugation energy[72], in the vibrational motion and finally in the correlation energy. Some or all of these terms may affect the energy balance sheet quite seriously. This is really the basis for the complaint[73] that the resonance energy of benzene may be anything from 10 to 70 kcal mol^{-1}. Recent efforts[33] to establish a rigorous theory of the chemical bond may help to rectify this situation.

To see how difficult this situation really is, let us do the simplest possible calculation of the energy of the hypothetical molecule (**20**) with the ethylene double bonds and three paraffin carbon-carbon single bonds in slightly different ways, (table 3). In method 1, we take the carbon-hydrogen bond energy of methane (99.3 kcal mol^{-1})[69] and using this same bond energy in ethane and ethylene we deduce the bond energies of the carbon-carbon bonds. We finally derive a resonance energy of 64 kcal mol^{-1} for benzene.

In method 2 we begin with the carbon-hydrogen bond energy appropriate to a large paraffin (98.7 kcal mol^{-1})[69]. Then we get a resonance energy of 50 kcal mol^{-1}. Then again we use Klages' method[70] translated into bond energies and atomisation energies, either with an ethylene double bond in method (a) or with the more stable *cis*-disubstituted ethylene double bond in method (b). The last method gives the familiar 36 kcal mol^{-1} for the resonance energy of benzene.

Table 3. *Simple calculations of the resonance energy (R.E.) of benzene*

	C-H bond energy	C-C bond energy	C=C bond energy	ΔH_a^o cyclohexatriene (**20**)	R.E.
Method 1	99.3 x 6	78.8 x 3	140.46 x 3	1253.6	64
Method 2	98.7 x 6	82.4 x 3	142.86 x 3	1268.0	50
Klages'(a)	98.5 x 6	83.1 x 3	143.7 x 3	1271.3	47
Klages'(b)	98.5 x 6	83.1 x 3	147.5 x 3	1282.4	36

Atomisation energy ΔH_a^o of ethane = 674.61; that of ethylene = 537.66. Data for methods 1 and 2 from ref. 69. All in kcal mol^{-1}. Klages values are from ref. 70 and are corrected to atomisation values. Klages (a) uses the C=C appropriate to ethylene and Klages (b) the stronger C=C of a *cis*-disubstituted ethylene in a six-membered ring.

The significant point is that a difference of about 30 kcal mol^{-1} in the resonance energy has emerged at the end of the calculations simply because of minor and arbitrary decisions made in the early stages of this simple calculation; there are many other decisions which could reasonably have been made at the different steps of the calculation. This general situation should always be borne in mind when one is confronted with elaborate computations which purport to yield accurate numerical values for resonance energies. Indeed, at this stage we are merely interpreting the experimental data in order to derive a number to compare with the theoretical computations! This general situation has led some workers to discard the whole idea of resonance energies of hypothetical molecules. It seems to us, however, that a more constructive approach is to attempt more accurately to understand the factors which make up bond energies so that we can work out the total energies of hypothetical molecules with some confidence. This point is further considered in chapter 3.6.

Table 4. *Heats of hydrogenation, ΔH_{H_2} of some unsaturated compounds.*

Molecule	$-\Delta H_{H_2}$ kcal mol^{-1}	Conditions	Resonance Energy kcal mol^{-1}
ethylene (a)	32.82	gas phase, 82°	–
propene (b)	30.12	"	–
2-methylpropene (b)	28.39	"	–
cis-but-2-ene (b)	28.57	"	–
cyclopentene (c)	26.92	"	–
cyclohexene (d)	28.59	"	–
cycloheptene (e)	26.52	"	–
cyclopentadiene (f)	50.87	"	–
pyrrole (g)	31.6	not stated	18.5
thiophen (g)	35.6	"	17.1
furan (g)[1]	37.2	"	14.9
benzene (f)	49.8	gas phase, 82°	36.0
naphthalene (h)[2]	80.0	gas phase	not stated
tropylidene (i)	70.49	soln., 25°	9.0
heptafulvene (j)	92.62	"	13.2
azulene (j)[3]	99.0	"	28.3
tropone (j)[4]	67.58	"	11.9

1. Compare with the data of ref. k.
2. Reduced to *trans*-decalin
3. Resonance energy includes solvation energy.
4. Reduced to cycloheptanone.

References for Table 4

a. Kistiakowsky, G.B., Romeyn, Jr., H., Ruhoff, J.R., Smith, H.A. and Vaughan, W.E. (1935) *J. Amer. Chem. Soc.*, **57**, 65

b. Kistiakowsky, G.B., Ruhoff, J.R., Smith, H.A. and Vaughan, W.E. (1935) *J. Amer. Chem. Soc.*, **57**, 876

c. Dolliver, M.A., Gresham, T.L., Kistiakowsky, G.B. and Vaughan, W.E. (1937), *J. Amer. Chem. Soc.*, **59**, 831

d. Kistiakowsky, G.B., Ruhoff, J.R., Smith, H.A. and Vaughan, W.E. (1936) *J. Amer. Chem Soc.*, **58**, 137

e. Conn, J.B., Kistiakowsky, G.B. and Smith, E.A. (1939) *J. Amer. Chem. Soc.*, **61**, 1868

f. Kistiakowsky, G.B., Ruhoff, J.R., Smith, H.A. and Vaughan, W.E. (1936) *J. Amer. Chem. Soc.*, **58**, 146

g. Quoted in ref. 69, p 81

h. Quoted in ref. 69, p 75

i. Turner, R.B., Meador, W.R., Winkler, R.E. (1957) *J. Amer. Chem. Soc.*, **79**, 4116

j. Turner, R.B., Meadow, W.R., Doering, W. von E., Knox, L.H., Mayer, J.R. and Wiley, D.W. (1957) *J. Amer. Chem. Soc.*, **79**, 4127.

k. Dolliver, M.A., Gresham, T.L., Kistiakowsky, G.B., Smith E.A. and Vaughan, W.E. (1938) *J. Amer. Chem. Soc.*, **60**, 440

We turn now to the hydrogenation experiment and its interpretation.[74] Tables of heats of hydrogenation are available[69] and some examples are given in table 4. Resonance energies can be deduced from heats of hydrogenation as follows. The heat of hydrogenation of cyclohexene,[69] used as a more realistic standard than ethylene for dealing with benzene, is -28.59 kcal mol^{-1}. Three times this is -85.8 kcal mol^{-1} and this is supposed to represent the heat of hydrogenation of the hypothetical molecule cyclohexatriene (**20**). The observed heat of hydrogenation of benzene is -49.8 kcal mol^{-1}, so benzene is apparently 36 kcal mol^{-1} more stable than is cyclohexatriene (**20**). Had we chosen ethylene as standard, the value of 36 would have become 48.7, showing again how sensitive is the final value to the assumptions made at the beginning. The energetics of the cyclohexene work are depicted in figure 14.

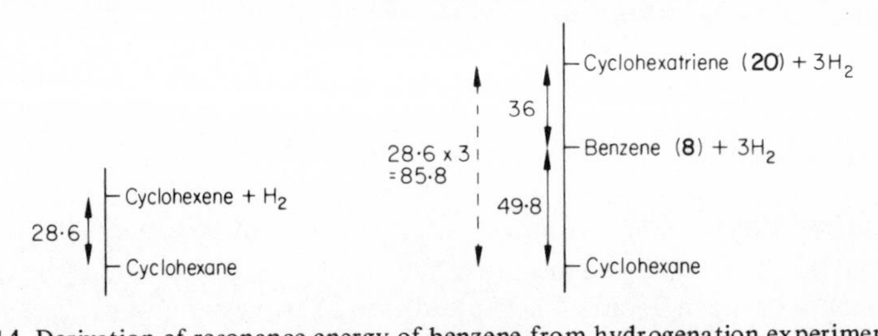

Figure 14. Derivation of resonance energy of benzene from hydrogenation experiments (values in kcal mol^{-1})

It would be better to call the resonance energy the 'stabilisation energy' as discussed in chapter 3.6, but the term resonance energy is now so thoroughly embedded in the literature that it is probably less confusing to continue to use this latter term.

At first sight, this hydrogenation result seems straightforward and attractively simple but this simplicity is largely deceptive. In reality, we face virtually the same problems here as in the combustion experiment because we are now dealing with four organic molecules, cyclohexene, cyclohexane, cyclohexatriene and benzene itself. The situation is simplified by the fact that the total energy of cyclohexane cancels out of the energetics, but we must be able to write down the total energy of cyclohexene, cyclohexatriene and benzene before we can deduce the stabilisation energy. What is usually done is to assume that certain cancellations are valid. For example, the carbon-hydrogen bonds of all three molecules are assumed the same in the above simple calculation. But assumptions of this kind often depend on subjective judgements about the bonds being the 'same' in different molecules. Moreover, cancellation procedures only apply to certain cases so they lack generality.

The combustion and hydrogenation experiments meet with complications when we deal with heterocycles such as pyridine, thiophen *etc* because to get useful information from these experiments requires detailed knowledge[75] of the products of combustion and hydrogenation and this is not always available.

For anions and cations, hydrogenation experiments are possible[76] but the interpretation is complicated by the necessity of making accurate corrections for the heat of solution of the ions.

An interesting area is that of the strained aromatic molecules,[76a] but the additional feature of strain energy is outside our present brief and we will not generally discuss molecules of this type.

2.3.2 Equilibrium Energetics

Much of the experimental information in this area relates to acid-base equilibra.

A classical example of the effect of aromaticity on the acidity of a hydrocarbon is that of cyclopentadiene **(14)** and its anion **(16)** (figure 15).

CH_2 14 ⇌ 16 + H^+

Figure 15.

The acid dissociation constant (expressed as the pK_a) of cyclopentadiene[77,78] is about 15, while that of a saturated hydrocarbon is difficult to estimate[78] but is probably between 35 and 40. The addition of two ester groups[79], as in figure 16, stabilises the anion until it is water-soluble without decomposition.

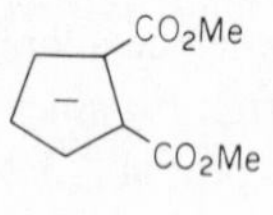

CO_2Me CO_2Me 21

Figure 16.

The general chemistry of the cyclopentadienyl anion and its derivatives has been discussed extensively.[9]

The above evidence is held to show that the 6π-electron, 5-carbon atom cycle of the cyclopentadienyl anion is a stabilised aromatic species. This is regarded as a major triumph of the (4n+2) rule.

Another celebrated example is the high basicity[80] of tropone (22) (*cf* chapter 5). This evidence is held to show that the 6π-electron 7-carbon cycle is aromatic.

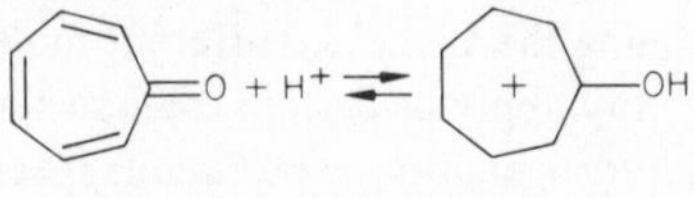

22 =O + H^+ ⇌ 23 —OH

Figure 17.

Another case[81,82] is that of the cyclopropenyl cation where the equilibrium shown in figure 18 is such that the pK_a of the triphenylcyclopropenyl cation, using ethanol as base, is similar to that of picric acid.[81a]

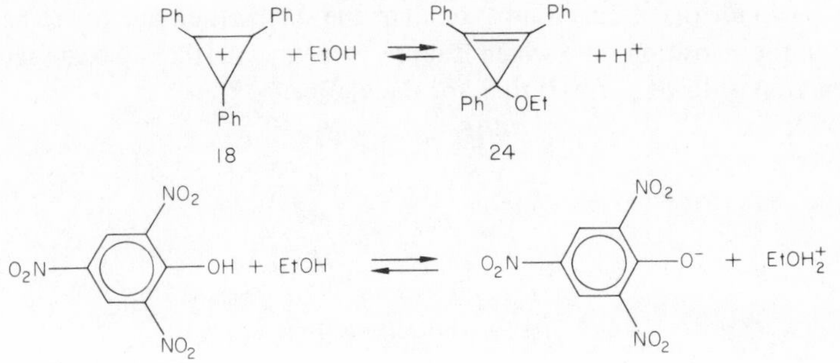

Figure 18.

This is held to show that the cyclopropenyl cation is particularly stable and hence is an aromatic species. This molecule does obey the $(4n+2)$ rule with n=o although there must be many other complications in a small ring system of this kind.

The only direct information given by the equilibrium experiment is the **difference** between the energy of the two organic species. Only if the total energy of one of the two is already known from other sources will the equilibrium experiment give useful information about the second species. In the above examples, we feel that we understand the energetics of the neutral molecule and so we can use the equilibrium experiment to get information about the ion.

Another difficulty with solution processes concerns the solvation energies which are usually large and often unmeasured or unmeasurable. A further difficulty arises when circumstances force us to compare four organic species, two pairs of acid-base equilibria. The hazards in so doing are self-evident.

To summarise this subsection we feel that the solvation energies of the ions should be known before useful information can be derived and we must be reasonably certain that we understand the energy of one of the two species before making inferences about the energy of the second species.

2.3.3 Chemical Reactivity Energetics

The oldest of all definitions of aromaticity is that aromatic molecules have the appearance of being unsaturated but are nevertheless chemically unlike olefines or acetylenes. Thus, olefines readily add molecular bromine to form a dibromide while aromatic systems react with molecular bromine to give the substitution product with retention of the aromatic type. In other words, the nature of the products is held to be a guide to the aromaticity of the reactant.

There are two interrelated aspects to this subject. One concerns the reaction rate and the mechanism of the reaction and the other the nature of the products. To illustrate the two points, figure 19 shows the bromination of an aromatic system ArH and of an olefine by molecular bromine. The free energies of the reactants we arbitrarily put equal. The free energy of activation for the ArH case will be greater than for the olefine if most or all of the stabilisation due to aromaticity is lost in the transition state which is represented by $(ArHBr)^+$. So the rate of the reaction is slower for ArH than for the olefine.

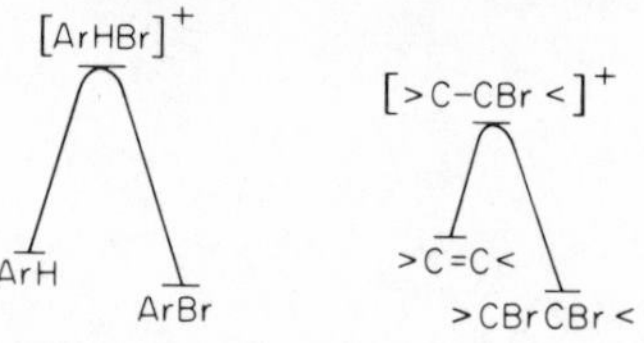

Figure 19. Vertical axis: Free energy.

Attempts to quantify these ideas have been made[83] but there are difficulties. In addition to selecting the correct reference olefine, one must be reasonably certain that the mechanisms of the two reactions are understood and are compatible. There is also the reservation that this definition must not be used when strongly activating or deactivating exocyclic substituents are present. Thus phenol is extremely reactive towards electrophiles and so would seem to be non-aromatic but one would surely wish to include the substituted benzenes together with benzene itself in the category of aromatic molecules. Yet another difficulty with this definition is that it cannot be applied to aromatic-seeming cations and anions such as the cyclopentadienyl anion.

Despite these shortcomings, the chemical reactivity definition seems to survive, perhaps because the nature of the products is so easily established experimentally while questions about mechanisms are more difficult to answer. But the advantage of thinking about the nature of the transition state is that one is closer to the fundamental electronic process of the chemical reaction. In passing, it is interesting to notice that the aromatic character of benzene asserts itself again after the transition state step. It does seem true to say that whatever the merits of the definition, the simple rule-of-thumb that olefines add bromine and uncharged aromatic rings tend to substitute with bromine is widely used.

In summary, it is tempting to say that if deductions from equilibrium processes can be unreliable then deductions from chemical reactivity are downright dangerous. Yet for many years much progress has been made in classifying molecules as aromatic using this shaky criterion. It seems that the use of this criterion is largely a matter of luck because it was developed using classical aromatic molecules and that transition states are indeed rarely aromatic species. We will continue to use this criterion of aromaticity bearing in mind how easily it may break down (*cf* chapters 4 & 5).

One point concerning chemical reactivity needs to be stressed. Taking the tropylium cation as an example, it is often said[84] that if reaction occurs equally at different carbon atoms, then it follows that the tropylium cation is the highly symmetrical species with sevenfold symmetry (**6**). This conclusion does **not** follow from the evidence. It may well be true that the ion consists of one sevenfold symmetric species but there could equally well be a rapidly equilibrating mixture of seven ions each with the classical structure (**25**) shown in figure 20. This erroneous argument, in which the products of a chemical reaction are held to provide information about small details of intermediates and transition states, has been used for many years in chemistry and it is remarkable how widely it is still used today.

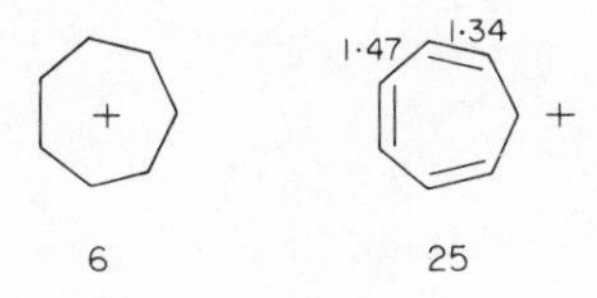

Figure 20.

2.4 Spectroscopic and Related Evidence

This material is divided into subsections related to the different types of spectroscopy. In addition, we have included some material which is not strictly spectroscopic but is closely related to it. Equally, a number of spectroscopic and related methods have been left out, such as X-ray, electron and neutron diffraction, microwave, mass spectroscopy and Mössbauer spectroscopy. These are omitted because, although they may give indirect evidence of aromaticity, *for example* in terms of molecular geometries, they themselves do not seem to tell us anything about aromaticity in a direct way.

2.4.1 Nuclear Magnetic Resonance Spectroscopy

This technique[35,85] consists of applying a static magnetic field to nuclei in order to separate the Zeeman levels of the nuclear spin angular momentum and then causing transitions to occur between these Zeeman levels by changing the z component of the spin angular momentum. This is done by putting in radio frequency power so that the magnetic vector of the radio frequency intereacts with the nuclear spin to change its z component.[35]

The field which a nucleus experiences consists of the applied external field, H_0, plus any magnetic fields which are already present in the molecule or are created in the molecule by the applied field H_0. These created fields are often envisaged as coming from the simple circulation of the electrons within the molecule (diamagnetic effect) and this is further broken up into contributions from the electrons on the nucleus in question, from the electrons of other atoms in the molecule and from any electrons which are free to move from atom to atom within the molecule. It is this last effect which generates the ring current in benzene. It may also happen that the electron clouds are distorted by the external magnetic field H_0. This is the paramagnetic effect.

The ring current idea is held to explain why the benzene protons and the ring protons of its homologues resonate at low applied field as compared with the non-aromatic protons (*cf* table 5).

Table 5. *Shift values of protons in aromatic and non-aromatic systems*

Molecule	τ Value
ethylene (a)	4.66
cyclo-octatetraene (b)	4.31
benzene (b)	2.734
naphthalene (c)	2.33 (α), 2.69 (β)
anthracene (c)	2.01 (α), 2.56 (β)
[14] annulene (d)	2.4 (outer), 10.0 (inner)
[16] annulene (e)	4.8 " -0.3 "
[18] annulene (d)	0.7 " 13.0 "
thiophen (f)	2.65 (α), 2.87 (β)
furan (f)	2.54 (α), 3.59 (β)
pyridine (g)	1.42 (2), 2.89 (3), 2.47 (4)
$C_5H_5^-$ (g)	4.33
$C_7H_7^+$(g)	0.82
cyclopentene (h)	4.40 (vinyl), 7.72 (allyl)
cyclohexene (c)	4.42 (olefinic)

References for Table 5

a. Lynden-Bell, R.M. and Sheppard, N. (1962) *Proc. Roy. Soc.* **269A,** 385

b. Tiers, G.V.D. (1958) *J. Phys. Chem.*, **62**, 1151

c. Ref. 99

d. Ref. 59

e. Schröder, G. and Oth, J.F.M. (1966) *Tetrahedron Letters,* 4083

f. Abraham, R.J. and Thomas, W.A. (1966) *J. Chem. Soc.* **(B).** 127

g. Schaefer, T. and Schneider, W.G. (1963) *Canada. J. Chem.*, **41**, 966 based on benzene τ = 2.734

h. Wiberg, K.B. and Nist, B.J. (1961) *J. Amer. Chem. Soc.*, **83,** 1226

One deduction from this general picture is that if we have a molecule with a proton above the ring, the ring current is expected to change only slightly the position of the proton signal from that of an 'aliphatic' proton and this expectation has been confirmed.[86] Also, if there is a proton inside a ring which can carry a diamagnetic ring current, then this proton should resonate at a high field. A clear example of this result is [*18*] annulene (**26**) which shows[59,87] the $A_{12}X_6$ spectrum at low temperature with τ = 0.72 for the twelve outer protons and τ = 12.99 for the six inner protons. Similarly an $A_{10}X_4$ spectrum is obtained with [*14*] annulene.[87]

Up to this point, the situation seems straightforward and attempts have been made to use the proton shifts to assign a degree of aromaticity to substituted benzenes,[88] to heterocycles,[89] and to other compounds. Measurements of the Faraday effect also relate to ring currents.[90]

It turns out, however, that there are two complications with this

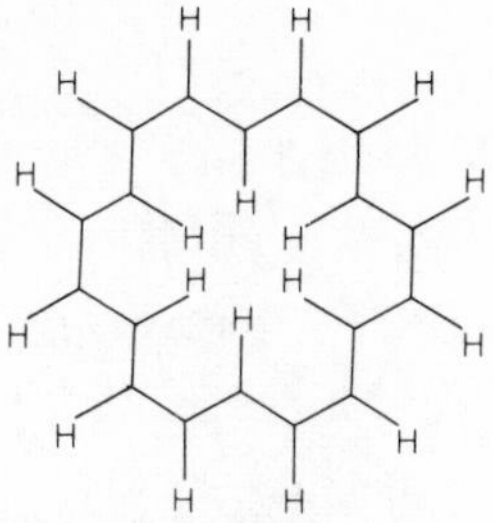

26

Figure 23. The π-electrons are not shown. .

attractive picture of the proton n.m.r. results. The annulene results (table 5) show the first complication. It is obvious that with [*16*] annulene and perhaps with [*12*] annulene the inner and outer protons resonate in the opposite sense from that which we would expect from our simple arguments. The accepted explanation of this result is that the [4*n*] annulenes have low-lying excited states leading to a large paramagnetic effect which dominates the situation, reversing the ring current and so reversing the positions of the ring proton signals. Some calculations[87,91,92] suggest that the ring current is sensitive to bond length alternation.

The second complication is that attempts to make the ring current concept into a quantitative one[36,93] have met with serious difficulties.[37,38,94] This situation suggests that there is at least one other effect similar in size to the ring current effect which operates in determining the position of the proton signal, and that care is required in interpreting n.m.r. data.

It should perhaps also be pointed out at this stage that there are serious experimental difficulties in such molecules as the annulenes whose n.m.r. spectrum varies markedly with temperature[39] and that it is often only at low temperatures that one gets unambiguous information. A particularly deceptive situation arises when, at higher temperatures, different conformers interchange more rapidly than the time scale of the n.m.r. experiment so that a very simple looking spectrum emerges from a complicated physical situation. One relevant case is that of the tropylium cation[95] which gives a single line at $\tau = 0.82$ (table 5). This looks like evidence for a highly symmetrical structure **(6)** (figure 20), but in fact it may equally well point to a rapidly equilibrating mixture of the seven classical ions of the type **(25)** (figure 20).

It has been suggested[96,97] that the shifts in the n.m.r spectrum which occur on dilution give some insight into the aromaticity of the solute, but the complexity of solution effects and of the solvation forces are such that one must accept this idea with some reserve at the moment.

It is also widely felt that the various kinds of n.m.r. coupling constants can yield information about aromaticity. One suggestion[98] is that the size of the $^3J_{\mathrm{HCCH}}$ coupling constant of vicinal protons is a measure of aromaticity.

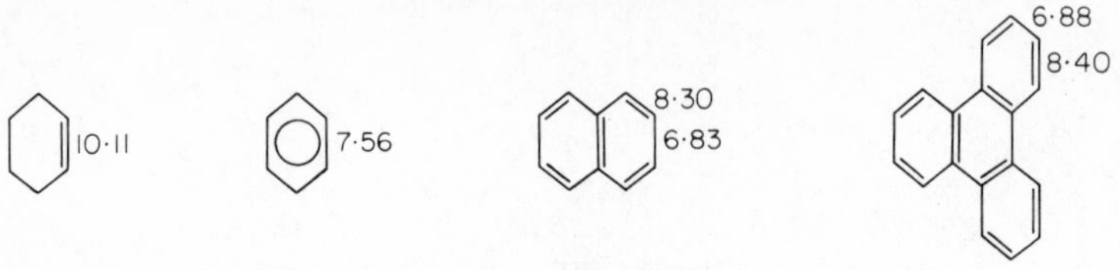

Figure 24. ^{3}J Coupling constants (in Hz) for some six-membered carbocyclic rings[99]

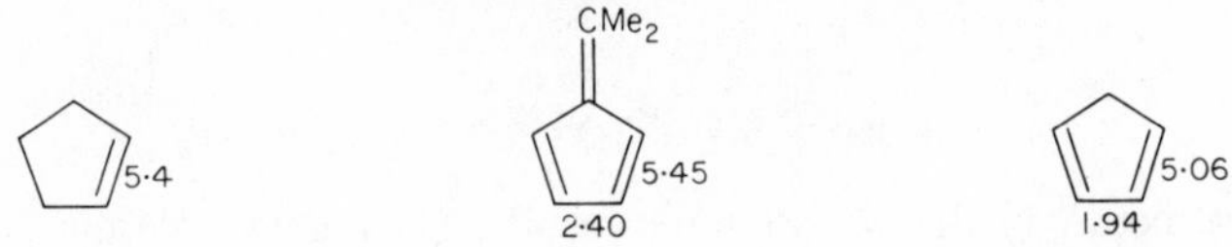

Figure 25 ^{3}J Coupling constants (in Hz) for some five-membered carbocyclic rings
(a) Ref. 98. (b) Smith, W.B. and Shoulders, B.A. (1964) *J.Am.Chem.Soc.* **86,** 3118.
(c) Manatt, S.L. and Elleman, D.D. quoted by Ganter C. and Roberts, J.D. (1966) *J.Am.Chem.Soc.* **88,** 741. (d) Ref. 180.(e) Ref. 161. (f) Ref. 224.

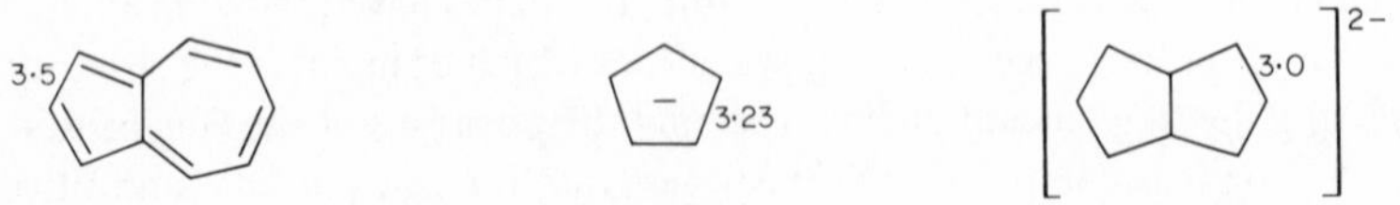

The examples given in figures 24 and 25 show that a large value of this coupling constant is connected with an olefinic bond and the smaller value with an aromatic bond. An important reservation is that the same type, *that is,* size and chemical type, of cycle must be compared among themselves. For example, a five-membered ring must not be compared with a six-membered ring and the presence of heteroatoms adds complications.

It might of course be argued that the ^3J coupling constants simply measure[99] bond lengths rather than directly measuring aromaticity. There is certainly a linear relationship between coupling constants and bond order[100] so we expect an inverse linear relation between coupling constants and bond length. If this is true, it is the bond length which remains as the fundamental criterion of aromaticity (*cf* chapter 2.2). The use of the directly bonded $J_{^{13}C\text{–}H}$ coupling constant as a measure of diamagnetic anisotropy and so of aromaticity has also been recommended,[101] since there is said to be a relationship between this coupling constant and the chemical shift of the proton.

2.4.2 Magnetic Susceptibility and Anisotropy

This type of evidence is not spectroscopic but it is closely related to the dynamic n.m.r. experiment and it is convenient to include it here.

The basic experimental fact is that molecules[102] respond differently to a magnetic field along different directions. That is, the magnetic susceptibility of a molecule is anisotropic and the complete statement of a molecule's susceptibility requires the values of the three principal components, χ_1, χ_2, χ_3 of the susceptibility tensor together with a statement of the three principal directions along which these compoents lie ($|\chi_3| > |\chi_2|$ or $|\chi_1|$). In simple molecules, these axes are often the familiar x, y, z axes, chosen in a natural way, but this is not always the case. The magnetic susceptibility as commonly measured χ_m(obs) is the average of these three components and is defined by

$$\chi_m(\text{obs}) = \frac{1}{3}(\chi_1 + \chi_2 + \chi_3)$$

When dealing with flat aromatic molecules, it is generally true that the two inplane components are about equal so we may define the anisotropy, ΔK (obs), by writing

$$\Delta K\,(\text{obs}) = \chi_3 - \frac{1}{2}(\chi_1 + \chi_2)$$

Some examples are given in table 6. Complete details are given in a standard text[102], and further values are tabulated.[102a]

Table 6. *Diamagnetic susceptibility, χ_m, diamagnetic susceptability exaltation, Λ, and diamagnetic anisotropy, ΔK, data, for some hydrocarbons**

Molecule	Susceptibility χ_m (obs)	Exaltation Λ	Anisotropy ΔK
cyclopentene	49.5	2.5	–
cyclopentadiene	44.5	6.5	34.2^c
cyclohexene	57.5	– 0.8	–
cyclo-octatetraene	53.9	– 0.9	–
tropylidene	59.8	8.1	–
benzene	54.8	13.7	58^a
naphthalene	91.9	30.5	119.8^b
anthracene	130.3	48.6	182.6^b
1,6-methano[10] annulene	112	36.8	–
heptalene	72	– 6	–
[16] annulene	105	– 5	–

All values are in units of -10^{-6} cm ^{3} mol^{-1}.
*The susceptibilities and exaltations are from ref. 104. The anisotropies are from (a) Le Fevre, R.J.W. and Murthy, D.S.N. (1969) *Aust. J. Chem.* **22**, 1415; (b) Lasheen, M.A. (1964) *Phil. Trans.* **256A,** 357; (c) Benson R.C. and Flygare, W.H., (1970) *J. Amer. Chem. Soc.*, **93,** 7523.

It has been appreciated for some time[103] that, since aromatic molecules have marked anisotropies which may be due to their having large ring currents, the anisotropy may well be a useful measure of their aromaticity. It is true of course that non-aromatic molecules may have significant anisotropies so allowance must be made for local contributions with some kind of group additivity scheme[103a] Thus, the experimental value of the anisotropy of cyclopentadiene is 34 and the calculated value from the group additivities is 30 (units as in table 6). In benzene, on the other hand, the two values are 58 and 16 respectively.

An alternative approach[104] is to compare the susceptibility χ_m(obs) with a calculated value of this quantity which is derived on some simple additivity assumption plus small correction terms. The outcome for an aromatic molecule is that, because of the ring current, χ_3 is large, and χ_m (obs) is larger than that calculated for the non-aromatic reference molecule. The difference is called the susceptibility exaltation, Λ, defined by

$$\Lambda = \chi_m\,(\text{obs}) - \chi_m(\text{calc})$$

It transpires that, for a classical aromatic molecule such as naphthalene, Λ is indeed large and for a non-aromatic molecule, Λ is about zero. Some exaltation values are given in table 6. Further discussion of individual cases is given in chapters 4 & 5.

Broadly speaking, this measure of aromaticity does seem to be a useful one and it is interesting to notice that it is covered by the definition of 'like benzene', since whatever feature of the electron organisation gives rise to this anisotropy in benzene also gives rise to the anisotropy in the other aromatic molecules. For paramagnetic molecules such as [*16*] annulene, Λ should be opposite in sign to that of benzene and, although the Λ of such molecules is indeed negative, it is rather smaller than might have been expected.[104]

2.4.3. Electron Spin Resonance Spectroscopy and the Aromaticity of Radical Ions

In this section on radical ions[105] we meet the interesting question of whether or not a radical can be aromatic. On some definitions of aromaticity, a radical cannot be aromatic. Thus, if the requirement of a closed shell of electrons,[106] whether in bonding molecular orbitals or not, is part of the definition of aromaticity then no radical can be aromatic. One might avoid this conclusion by saying that the extra electron of the benzene radical anion, for example, is outside the aromatic sextet but it is not altogether clear what 'outside the sextet' means. The ($4n$+2) rule is not obeyed in simple radicals which usually have an odd number of electrons. The chemical reactivity of a radical is certainly nothing like that of benzene.[107] So we might say that a radical is not aromatic.

On the other hand, it is quite likely that the carbon-carbon bond lengths of the benzene radical anion are close to those of benzene itself. This seems reasonable since the added electron of the radical anion is only weakly bound and is in a large orbit, so it plays little direct part in the bonding between the atoms which is still controlled by the same electrons as are present in the neutral benzene molecule. The present definition of aromatic as 'having an electron organisation like that of benzene' does permit such radicals as the benzene radical anion to be aromatic and this is a strong point in its favour. It does seem to be generally felt that the term 'aromatic hydrocarbon radical' is appropriate to such radicals together with the implication that the radical is an aromatic entity. To be conservative, perhaps one should say 'radical derived from an aromatic hydrocarbon' but this is clumsy terminology

One point which is established[108] quite clearly by the e.s.r. experiment with such radicals as the benzene radical anion, is that the added electron is spread over the molecule, very much as the first antibonding Hückel molecular orbital predicts it to be. Take careful note that we cannot deduce from this that the original 6 π-electrons are also in Hückel type molecular orbitals: they could be in chemical bonds despite the fact that the added electron is in a delocalised molecular orbital. Nevertheless, the fact that the extra electron is approximately described as inhabiting the antibonding Hückel orbital is perhaps the best evidence for the general validity of the Hückel type theories.

It is evident that the radical cation is a very different matter from the radical anion in a π-electron system, if, as is usual[109], the amount of energy involved in removing an electron is some 10eV. Clearly, if one of the six π-electrons of benzene is removed, the 'aromatic sextet' is disrupted so perhaps the aromaticity will not survive this process. This seems to us to be an interesting area for exploration.

2.4.4 Vibration Spectroscopy

The vibration spectrum of a molecule, both infrared and Raman, is an easily available quantity, but to analyse the spectrum completely and so derive a set of force constants is a formidable task.[110] Even if this task is completed, the results do not tell one in any clear way whether the electron organisation in the molecule is that characteristic of aromaticity. It might be possible to use the force constants from, say, ethylene and ethane in aromatic hydrocarbons and then show that the resulting calculated vibrational spectrum of the aromatic molecule does not agree with the observed spectrum, but this is an indirect and negative piece of evidence. The systematic use of vibrational spectra to detect aromaticity does not seem promising.

It is of course true that the vibrational spectrum is relevant to the geometry of the molecule and, in special cases, this can indirectly point to the aromaticity of a molecule. The examples are usually of high symmetry, such as the tropylium cation,[111] where the infrared spectrum is very simple, showing only four lines of reasonable intensity, which suggests a highly symmetrical geometry with many forbidden transitions. It seems likely that this cation does indeed have D_{7h} symmetry (*cf* chapter 5.2). This method is obviously only suited to a few special cases of high symmetry and the general conclusion is still that vibrational spectroscopy is not helpful with the aromaticity problem at this time.

2.4.5 Electronic Spectroscopy

It has sometimes been suggested that the position of the ultraviolet or visible maximum gives an indication of the aromaticity of the molecule.[112]

This approach is hedged around with difficulties. Even if all the experimental problems concerning medium effects can be eliminated, we are again faced with the difficulty that two molecules, the ground state and the excited state, are intimately involved in the experiment and we observe essentially a difference between two molecules which differ drastically in their electron organisation. And the task of interpreting ultraviolet spectra is formidable because only in quite small molecules is the resolution sufficient to allow a detailed rotational analysis to establish even the geometry of the excited state.[113] In all other cases, one has to guess this geometry and it may or may not be the same as that of the ground state.[113] In this circumstance it is obviously difficulty to make firm statements about the ground state from the observation of the ultraviolet spectrum.

Perhaps the clearest hint that one gets from the ultraviolet spectrum concerning the aromaticity of a molecule concerns the case of a highly symmetrical molecule. Then many of the transitions are forbidden and one can perhaps infer that the high symmetry is a consequence of the aromaticity. But even here one is using the ultraviolet spectrum to get geometrical information which in turn is related to the aromaticity.

2.4.6 Ionisation Potentials and Photoelectron Spectroscopy

The measurement of the ionisation potential is at first sight an attractive means of deciding whether a molecule is or is not aromatic. The attractive feature of this method is that it seems to give *directly* the electron energy levels of the molecule because, at least in the closed-shell case, the ionisation potential is the eigenvalue or orbital energy of the molecular orbital equation (**14**) of chapter 3.4.[114] Experimentally, the ionisation potential is determined by photo-ionisation[115], by direct measurement of the electron impact [116] and, most elegantly, by photoelectron spectroscopy.[117] In this last experiment a photon of known energy, $h\nu$ is fired at the molecule, A, and the energetics of the process

$$A + h\nu \longrightarrow A^{+} + e^{-}$$

are established by measuring the kinetic energy of the ejected electron. The method has the advantage that states which are not easily observed by other methods are seen in this experiment.

It has now become customary to say that **the** electron energy levels are being measured in the photoelectron experiment. This raises a delicate point because, as discussed in the theory section there is no such thing as **the** electron energy levels of a molecule. Only the total energy of a molecule is a well defined quantity, not the energy of the individual electrons. What the photoelectron experiment really measures is the *difference* between the total energy of the neutral molecule and the total energy of the positive ion. The positive ion may of course be in various states. To express this point in another way, the molecular orbitals are not unique but really act as building blocks for the unique many-electron wave function and it is possible to use a variety of different types of building block to give the same final edifice. So the electron energy levels as determined by photoelectron spectroscopy may not have relevance for ground states. The detailed discussion of the methane molecule[33] shows this quite clearly (*cf* figure 26).

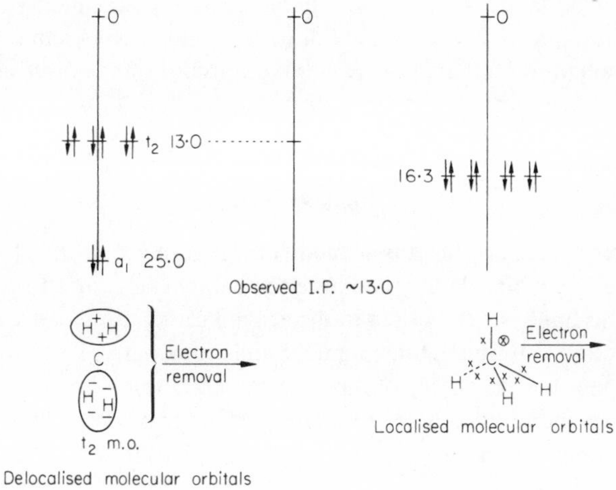

Figure 26. Ionisation potentials[119] of methane in terms of delocalised and localised molecular orbitals (vertical axes represent energy in eV)

Once again we find ourselves in the position that the information we have refers to two species, in this case the neutral molecule and the positive ion, and not directly to one species. It seems that the familiar canonical molecular orbitals (the delocalised a_1 and t_2 molecular orbital of methane) do indeed describe the positive ion reasonably well and that there is no extensive rearrangement of the other electrons when an electron is removed from the canonical molecular orbital. Notice that the Koopmans' theorem,[118] that the ionisation potential is the eigenvalue of the Hartree-Fock equation, will apply to any set of molecular orbitals and not just to the canonical ones.

The critical test of whether we should use the canonical molecular orbitals in discussing ionisation potentials comes when one compared the calculated ionisation potential and the experimental ionisation potential using various sets of molecular orbitals.[119] Such a comparison is difficult with large aromatic molecules but it has been done with the small molecules with convincing results.[119]

An experimental result which gives a clear warning about the dangers of too simple an interpretation of ionisation potentials is the fact[120] that the ionisation potential of cyclo-octatetraene is 8.04 eV while that of ethylene is 10.5 eV. It does not follow from this result that in the neutral cyclo-octatetraene molecule there is extensive interaction between the double bonds; rather it follows that the positive ion is **not** formed simply by removing an electron from one of the double bonds. Evidently this ion is formed by removing an electron partly from some or all of the four double bonds.

The direct relevance of ionisation potentials to aromaticity is accordingly quite limited and complications also arise from the fact that there are often substantial errors (± 0.5 eV) in ionisation potential measurements. This is a large amount of energy by the standards of resonance energies and this means that ionisation potentials are not likely to be a useful source of information about the general question of aromaticity.

2.5 Dipole Moments

It might seem that, since the dipole moment is a property of the electronic ground state of a molecule, it could have some direct relevance to the aromaticity question. This hope is not realised in the general case because there seems to be little connection between dipole moments and aromaticity. Part of the difficulty here is our poor understanding of dipole moments. Even in small molecules such as carbon monoxide, there are several distinct major contributors[121] to the final dipole moment as shown in figure 27. This result is a clear

3·26 0·09 0·79 2·13
←+C+→ ←+O+→

μ_q = 0·54 (Atomic charge contribution)
μ_T = 1·29 (Total calculated dipole moment)
μ_{expt} = 0·11

Figure 27. Contributions[121] (in Debye units) to the dipole moment of carbon monoxide

warning that naive interpretations of the magnitude and sign of the dipole moments in terms of semi-empirical computations can be misleading.

There are a few special cases where it is certainly tempting to say that the dipole moments do have some bearing on the aromaticity question. These are molecules in which the electron rearrangement from the aromatic to the non-aromatic structure would be accompanied by large charge migrations. An example is fulvene (**32**), where the aromatic structure (**33**) will have a large π-electron dipole moment – something like 10 Debye units. But it is certain that much of this large dipole moment would be offset by the back polarisation of the σ-electrons thus partly evening-out the charge

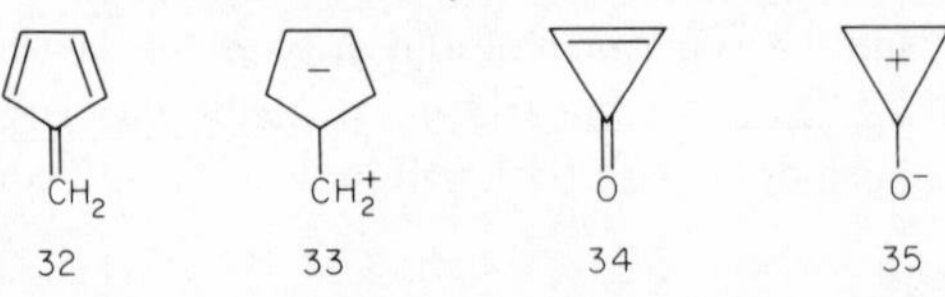

Figure 28.

distribution. Experience[121] suggests that a value for the total dipole moment of **(33)** of 2 to 4 *D* might be correct. The dipole moment of fulvene[122] is 1.1 *D* so there could be some 25 to 50 per cent of structure **(33)** in the ground state of the molecule (*cf* chapter 4). However it must be remembered that cyclopentadiene **(14)** itself has a dipole moment[123] of 0.42 *D* and cyclohexa-1,3-diene has a dipole moment[124] of 0.44 *D* and, since the origin of these two dipole moments is uncertain, we must accept that hydrocarbons of this general type may have dipole moments of about 0.5 *D* whose origin is obscure. So we cannot attribute much significance to small dipole moments at this time.

When the dipole moments are larger, the arguments seem safer. As an example cyclopropenone **(34)** with alkyl or aryl substituents has a dipole moment of about 5.0 *D*[125] while that of acetone is 2.9 *D*[126] The extra 2.0 *D* may then come from the presence of structure **(35)** in the ground state of the cyclopropenone molecule. But even here the conversion of carbon-hydrogen to carbon-carbon bonds or the back polarising of the σ-bonds may easily explain the 2.0 *D* discrepancy and only a full computation of the molecular wave function can explain the result.

Generally speaking, then, the interpretation of dipole moments is a difficult problem[127] and it seems doubtful that much information about the aromaticity of molecules can be derived from this source.

2.6 Miscellaneous Experimental Information

The nuclear quadrupole experiment[128] measures the electric field gradient at the quadrupolar nucleus and is a potentially valuable source of information about the wave function of the molecule because the electric field is determined by the electron organisation. There is the disadvantage that few common nuclei have quadrupole moments and that neither carbon nor hydrogen is among them. The most promising area of direct interest to the aromaticity question is in the nitrogen heterocycles such as pyridine, where the quadrupolar ^{14}N nucleus is part of the aromatic ring. Substituents which have a quadrupole moment, such as chlorine, may be attached to any aromatic molecule, but then the evidence relating to the aromaticity of the ring is less direct. It seems likely that important developments will take place in this area shortly.

The cracking pattern from the mass spectroscopy experiment seems to have little direct relevance to the question of the aromaticity of the molecule. An interesting exception to this generalisation is the rearrangement of the benzyl cation to the tropylium cation which was demonstrated in the mass spectrometer[129] (*cf* chapter 5.2)

The electron affinity of a molecule[130] measured, for example, by reductive polarography, is the energy required to create the anion radical. The electron affinity does relate to the calculated energy levels of the molecule in a general way, but the information which is obtained seems to relate more to the added electron than to the electrons which are already present. So the electron affinity may offer valuable confirmation of a general theory but it is not necessarily relevant to the aromaticity of the neutral molecule.

2.7 Summary and Comments

From the foregoing material we conclude that there are three main sources of information about the aromaticity of a molecule. The first is the geometry of the molecule, where both the planarity and the bond length give a clear guide to whether the molecule is 'like benzene'. The second is the atomisation energy of the molecule, however this is measured and expressed. The third is the n.m.r. spectrum and the related diamagnetic anisotropy, commonly expressed in terms of the exaltation. The various other properties which have been mentioned do on occasion give some insight into aromaticity. An obvious example of this point is chemical reactivity and, on occasions, we do have to rely on these less straightforward pieces of evidence in order to get some idea of whether a molecule is aromatic.

It seems that only the above mentioned three pieces of evidence are really decisive in deciding whether a molecule is aromatic.

3. Theoretical Ideas on Aromaticity

Introduction – qualitative molecular orbital theory – the ideal situation – the current situation – lower approximations and σ–π separability – resonance and stabilisation energies – CNDO theory– theories with σ-electrons treated as simple two-electron chemical bonds – theories in which the σ-electrons are neglected – summary

3.1 Introduction

If we define 'aromatic' as 'having an electron organisation like that of benzene' the first task is to say what the electron organisation of benzene really is. To do this, we must discuss the difficult subject of wave function computation in some detail. Indeed, we have elaborated this subject to a much greater extent than is usual in an organic text. We give a few examples in each section but we make no attempt to give the kind of detailed numerical information which is already available in the theory texts.

The first point which must be appreciated is that although the π-electron* molecular orbital description of the benzene wave function is so familiar that it seems to have the status of a basic truth, the position is that no really accurate computation of the wave function of a molecule as large and complex as benzene is even remotely possible. By 'really accurate' is meant of *known* accuracy to within one to two kcal mol^{-1} in the total energy. The most accurate computations so far done on benzene[131] and similar molecules[132] still give errors which are about three powers of ten larger than this. Unfortunately, there is little real prospect of substantial improvement in this level of accuracy in the foreseeable future. In particular, we know very little of the valence shell σ-electrons which outnumber the π-electrons by about 4 to

*The term 'π-electron' should be confined, strictly speaking, to linear molecules such as acetylene whose π molecular orbitals have unit angular momentum about the internuclear axis. But the term is now extended to any double-bonded system.

1. We often suppose that these electrons are in simple chemical bonds and this may be true but we have no evidence on the point. For example, we often suppose that the carbon-carbon σ-bonds of benzene and of ethylene are essentially the same, but there is no substantial evidence for this assumption.

We are thus committed for the foreseeable future to using approximate computations of unknown reliability and nothing is gained, while a great deal is lost, by pretending otherwise. The art of using such methods lies in discarding all features which either singly or together, do not significantly affect the particular property of the particular molecule which we wish to calculate. Judicious choices in such crude computations can lead to numerical values which agree with experimental values and with the agreement being for the physically correct reason. A good example here is the one-electron energy quantities and other one-electron properties which can be calculated reasonably accurately with theories which give inaccurate values for bond energies.[133]

There are available several detailed chronological accounts of the theoretical material relevant to aromaticity.[1-18] In the present section, we have attempted to use a contrasting point of view. We begin by describing what seems to us to be the ideal situation in our theoretical understanding of aromaticity which may be achieved in some distant future. Then we show how far the present day situation has advanced towards this hypothetical ideal situation and finally we point out the historical steps which led to the present stage. This is an antichronological presentation which helps to throw light into some unfamiliar places. In particular, we stress[134] the importance of the σ-π interaction which was revealed by thorough going computations.

3.2 Qualitative Molecular Orbital Theory

The purpose of this section is to introduce the basic ideas of molecular orbital theory[135] and also to show the close relationship between this theory and the conventional chemical valence theory. The intention in this subsection is to present the underlying physical situation, divorced as far as possible from the mathematical complexity which often overlays and obscures it. All of this material is of course available in the standard texts on valence theory but it is not usually given in this format.

The molecular orbital idea is basically a physicist's rather than a chemist's point of view. The starting point of this approach is to imagine the nuclei of the molecule fixed in space, ignoring the vibrational motion altogether, and then add a single electron to this framework of bare nuclei. This electron will spread over the entire region of space occupied by the nuclei and the spread-out cloud forms a molecular orbital. This is the basic point of view which characterises the conventional molecular orbital theory.

The second step in this theory follows from the fact that there will be many such molecular orbitals and so, arguing by analogy with atomic structure theory where electrons are fed one after the other into the orbitals, the molecular orbitals are filled with pairs of electrons of opposite spin. This generates the qualitative picture of the molecular orbitals viewpoint.

The size and shape of the molecular orbitals will be given as a function of three dimensions and this function will be obtained by solving some kind of equation (see below). When this is done one gets a set of molecular orbitals each having a particular energy. Then the calculation of the wave function of the molecule is complete, although it still remains to compute values for the various properties of the molecule, such as the electric and magnetic dipole moments, magnetic susceptibilities, coupling constants and other electromagnetic properties. It is only in this second stage that we get values for observables of the molecule.

It may be remarked in passing that, although the above procedure sounds plausible, it contains assumptions of unpredictable accuracy. One such is the idea that the orbitals do not change much as the electrons are introduced. It is not clear that this is true because the forms of the orbitals are determined by the potential created by the electrons of the orbitals. But there is also the deeper underlying problem concerned with the uniqueness of the orbitals whose ramifications are extensive and are discussed later. For now we follow the naive argument used above.

On the question of determining the molecular orbitals, the basic mathematical machinery for describing electron motion was created by Schrödinger and others in the form of an equation whose solutions are the wave functions of the system together with the energy levels. We speak here of the many-electron wave function of the molecule but it is natural to expect that an analogous equation will apply to the individual molecular orbitals.

Since Schrödinger's equation is based on finding an operator corresponding to the total energy of the system and forming its eigenvalue equation, it is natural to suppose that we can find a quantity called the energy of the electron (or its molecular orbital) and set up a Schrödinger equation based on this energy quantity. There are difficulties here but putting these aside in this qualitative introduction, we might suppose that the energy of the electron will be its kinetic energy plus its potential energy of attraction to the nuclei and repulsion by the other electrons. The operator corresponding to the kinetic energy is $(-\hbar^2/2m)\nabla^2$ where

$$^2 = \frac{\delta^2}{\delta_x{}^2} + \frac{\delta^2}{\delta_y{}^2} + \frac{\delta^2}{\delta_z{}^2} \qquad (1)$$

Let us denote the potential energy of the electron situated at the point x_1, y_1, z_1 by the symbol $V(x_1, y_1, z_1)$ or simply V(1). Then the equation, which is not exactly Schrödinger's equation but is a close relative called the Hartree-Fock equation, will be

$$\left[\frac{-\hbar^2}{2m}\nabla^2(1) \quad +V(1)\right] \quad \phi_i(1) = e_i\,\phi_i(1) \qquad (2)$$

If one knows what V(1) is, then the solution of this equation is a mechanical exercise which is discussed in great detail in any standard text on molecular orbital theory. Briefly, the equation is converted from its operator-function form to the equivalent matrix-vector form which is simply the secular determinant and secular equations. The results are a set of energy levels or numerical values of e_i plus the forms of the molecular orbitals which are expressed in the linear combination of atomic orbitals form

$$\phi_i = \sum_p c_{ip} \nu_p \tag{3}$$

where the ν_p are the atomic orbitals and the c_{ip} are the linear coefficients. A primitive example of this procedure is the bonding molecular orbital of the hydrogen molecule where

$$\phi = (1s_a + 1s_b) / (2 + 2S)^{1/2} \tag{4}$$

and S is the overlap integral between the two atomic orbitals.

This procedure generates the 'canonical' molecular orbitals. The term 'canonical' arises because the solving of a secular equation is essentially a matter of diagonalising a matrix and the diagonal form of a matrix is referred to as the 'canonical' form. The corresponding molecular orbitals are called the canonical ones. These orbitals have the advantage of being mathematically well defined but it is **not** obvious that these are the best molecular orbitals in a physical sense. The localised molecular orbitals are not canonical in the above sense but they have distinct advantages for some purposes.

The essential problem with all of this work is that we do not know what V(1) is. One way out of this difficulty, called the semi-empirical method, is to give up any idea of finding an explicit expression for V(1) and then treat integrals in which it occurs as disposable parameters whose values are to be determined by experiment. If one rejects this approach, and there are good reasons for so doing, then one must try to formulate V(1) explicitly and this is the non empirical approach which is coming to be favoured today. In this method, V(1) is formulated by a series of well defined approximations.

The above is a summary of the basic physical ideas of conventional molecular orbital theory. This molecular orbital theory is now well established and its success is undeniable for many circumstances and many purposes. It suffers from one major defect, however, in that it seems to have no connection with the well-established and very successful chemical valence theory based essentially on the two-electron two-centre chemical bond. It is true that the chemical valence theory of the two-electron bond has lacked a rigorous quantum mechanical foundation until recently,[33] but it is surely inconceivable that this valence theory can be spurious when one thinks of its successes in the vast realm of the chemistry of saturated carbon compounds. It seems to us that this dichotomy has retarded the acceptance of molecular orbital theory in chemistry to a marked extent.

The problem arises because there exists a subtle complication in the entire approach to molecular orbital theory outlined in this section. This complication arises from the fact that electrons are by no means such simple things as we have assumed so far. To summarise the matter in a formal way it is enough to say that orbitals are not unique, but this concept is a difficult one for a physical scientist who intuitively supposes that the orbital is a given fixed physical 'thing' in the same sense that, say, the trajectory of a large physical object is a fixed definite thing. But electron functions differ from such things as classical trajectories in a way which originates in the antisymmetry principle which the many-electron wave function must obey. This somewhat abtruse point is dealt with at length in texts on quantum theory but the result is that, for practical purposes, the wave function must be set up as a determinant in order to ensure this antisymmetry. This in turn means that orbitals are not fixed and definite things because determinants are not changed by, say, adding a copy of one row to another row.

As an example, suppose that we have a four-electron, two-orbital (ϕ_a, ϕ_b) situation. The wave function is then

$$\Psi = | \phi_a(1)\ \bar{\phi}_a(2)\ \phi_b(3)\ \bar{\phi}_b(4) | \tag{5}$$

where the bar denotes a β spin. Suppose that we form the sum and difference of the two molecular orbitals thus

$$\chi_a = (\phi_a + \phi_b) / (2)^{1/2} \qquad \chi_b = (\phi_a - \phi_b)(2)^{1/2} \tag{6}$$

and construct the wave function

$$\Psi' = | \chi_a(1)\ \bar{\chi}_a(2)\ \chi_b(3)\ \bar{\chi}_b(4) | \tag{7}$$

The point is that the two wave functions Ψ and Ψ' are identical (not roughly equal but identical) although different orbitals appear in construction of the two determinants. It is clear then that there is no such thing as *the* molecular orbitals of a molecule. As an explicit example of this point, the methane molecule's wave function may be viewed either from the localised molecular orbital or from the delocalised molecular orbital point of view. The canonical molecular orbitals $\phi(a_1)$ and $\phi(t_2)$ are approximately:

$$\phi(a_1) = 0.130\,(1s_{Ha} + 1s_{Hb} + 1s_{Hc} + 1s_{Hd}) - 0.206\,(1s_C) + 0.698\,(2s_C)$$

$$\phi(t_{2x}) = 0.305\,(1s_{Ha} + 1s_{Hb} - 1s_{Hc} - 1s_{Hd}) + 0.585\,(2p_x)$$

and similarly for the ϕ_{2py} and ϕ_{2pz} molecular orbitals. The localised molecular orbitals[33] are the four molecular orbitals χ_a, χ_b, χ_c and χ_d which are equivalent. The form of χ_a is $\chi_a = 0.531\,(1s_{Ha}) - 0.098\,(1s_C) + 0.355\,(2s_C) + 0.501\,(2p_C) - 0.090\,(1s_{Hb} + 1s_{Hc} + 1s_{Hd})$.

This point has been made a number of times in the history of this subject[135a] but it does not seem to be widely appreciated by chemists generally.

A second example relevant to aromaticity is that of benzene itself. The occupied canonical molecular orbitals are determined by symmetry alone and have the forms (normalisation aside)

$$\phi(a_{2u}) = \pi_a + \pi_b + \pi_c + \pi_d + \pi_e + \pi_f$$

$$\phi(e_{1g}) = 2\pi_a + \pi_b + \pi_f - 2\pi_d - \pi_c - \pi_e$$

$$\phi(e_{1g}) = \pi_b + \pi_c - \pi_e - \pi_f$$

These molecular orbitals seem very different from the localised molecular orbitals (χ) of a Kekulé structure which are naturally written

$$\chi_{ab} = \pi_a + \pi_b \qquad \chi_{cd} = \pi_c + \pi_d \qquad \chi_{ef} = \pi_e + \pi_f$$

This difference is largely illusory however as one can see by applying a simple transformation to these three localised molecular orbitals to get the three delocalised but not canonical molecular orbitals (γ) which generate the same wave function as that of a Kekulé structure.

These are

$$\gamma(a_2'') = \pi_a + \pi_b + \pi_c + \pi_d + \pi_e + \pi_f$$

$$\gamma(e'') = 2\pi_a + 2\pi_b - \pi_c - \pi_d - \pi_e - \pi_f$$

$$\gamma(e'') = \pi_c + \pi_d - \pi_e - \pi_f$$

These molecular orbitals transform under D_{3h} not D_{6h} as the canonical molecular orbitals do. It is clear however that, when the molecular orbitals of the Kekulé structure are put into this form, the lowest energy molecular orbital is the same in both sets and even the doubly degenerate molecular orbitals are so similar that the distinction between the canonical molecular orbitals and the delocalised molecular orbitals of a Kekulé structure is a subtle one.

We have elaborated this point about the lack of uniqueness of the molecular orbitals at some length because it is the pivotal point connecting chemical valence theory with which most chemists are familiar and conventional molecular orbital theory which is less widely understood by chemists. An appreciation of this point will go far towards overcoming the hesitation which many chemists feel towards molecular orbital theory.

3.3 The Ideal Situation

The exact many-electron wave function[29, 135] of a molecule (Ψ exact) is defined as the solution of Schrödinger's equation (for an n-electron system)

$$H(1,2...n)\ \Psi_{\text{exact}}\ (1,2...n) = E_{\text{exact}}\ \Psi_{\text{exact}}(1,2...n) \qquad (8)$$

where E_{exact} is the total energy of the molecule excluding the nuclear repulsion and is given by

$$E_{\text{exact}} = \langle\Psi_{\text{exact}}|\ H(1,2..\ .n)|\ \Psi_{\text{exact}}\rangle \qquad (9)$$

where the total Hamiltonian, H, of the molecule is

$$H(1,2..\ .n) = \sum_{i=1}^{n} (-\tfrac{1}{2}\nabla_i^2) + \sum_{i=1}^{n}\sum_{a=1}^{k} (-Z_a/r_{ai}) + \sum\sum_{i>j=1}^{n} 1/r_{ij} \qquad (10)$$

The first term on the right of equation 10 represents the total kinetic energy of all the electrons, the second term represents the total potential energy of attraction between all the nuclei ($1, 2\ .\ .k$) and all the electrons and the third term represents the coulomb repulsion between all the electrons. Rigorous definitions are given in reference 136. The Born Oppenheimer separation is assumed to be valid[29] so that the nuclei are envisaged as being stationary. The wave function is a function of $4n$ variables ($x_1, y_1, z_1, x_2 . . . z_n$, plus the n spin variables $\sigma_1, \sigma_2 . . . \sigma_n$). The four variables of one electron are designated by the electron label.

If we knew this perfect wave function for a molecule, we might hope to single out some characteristic quality or property which we call the 'aromatic quality' of the molecule. It might even be possible to find a quantitative measure of the degree of aromaticity of the molecule. One of the simplest possible ways of doing this would be to evaluate the overlap integral between the benzene wave function and that of the molecule in question, or part of the wave function of the molecule in question to see how nearly identical are the two wave functions. Or some feature of the energy levels of the molecules may indicate aromaticity. There are many other possibilities and the essential point is that we should be able to distinguish within the wave function of a molecule some characteristic quality which we recognise as the aromatic quality or quantity of the molecule. The primitive example of this procedure which we have now is the idea of the aromatic sextet of the benzene wave function which reappears in a similar form in the cyclic $[C_5H_5]^-$ and $[C_7H_7]^+$ ions. The characteristic quality here is a cyclic structure with three low-energy doubly-filled molecular orbitals and whenever we see a molecular orbital pattern like this for the ground state of a molecule, we suspect that we are dealing with an example of aromaticity. This is the basis of the ($4n$+2) rule.[25]

It is obvious that the present mainstream of development in molecular wave functions points towards some extension of the self-consistent field, Hartree-Fock method of wave function computation.[136] In this method, we solve not equation (8) but the closely related equation (11).

$$H'(1,2\ldots n)\,\Psi_{HF}\,(1,2\ldots n) = E'\,\Psi_{HF}\,(1,2\ldots n) \qquad (11)$$

whose operator is

$$H'(1,2\ldots n) = \sum_{i=1}^{n}\left[-\tfrac{1}{2}\nabla_i^2 - \sum_{a=1}^{k} Za/r_{ai} \quad \sum_{j=1} (2J_j - K_j)\right] \qquad (12)$$

$$= \sum_{i=1}^{n} f(i)$$

and this operator is now a sum of one electron operators, $f(i)$, each of which is formed from the kinetic energy of one electron, the attraction of this electron for all the nuclei (1 . . .k) and the repulsion-exchange attraction of this electron for **all** the electrons. Notice carefully that E' is **not** a useful approximation for the E_{exact}, the latter is worked out using Ψ_{HF} with the *true* Hamilton H.

The operator H′ has the admirable quality of giving an exactly soluble eigenvalue equation for a molecular orbital as in equation (14) below. It is easy to show that this wave function Ψ_{HF} is a single determinant of doubly occupied molecular orbitals.

But even when we have solved equation (11), we are still faced with the problem that H' is not the correct operator, nor is Ψ_{HF} the accurate wave function of the molecule. An enormous amount of effort has gone into the computation both of the self-consistent field molecular orbitals[137] and into the question of the corrections which must be applied to this wave function in order to get the true wave function of the real molecule.[138] We have as yet little idea of what the final outcome will be.

The essential point which we must hope is true for our present purpose is that the 'aromatic quality' of a wave function is already revealed by the one determinant Hartree-Fock approximation. If it is not, then we have little chance of a genuine understanding of the concept of aromaticity for many years to come. In these circumstances, the only constructive assumption is that the Hartree-Fock method will indeed reveal at least the essence of the aromatic quality.

It should be noted that, once we have retreated to the Hartree- Fock level of approximation, we have assumed a number of things. The most important of these is the orbital approximation in which we suppose that an electron inhabits an orbital. This is not so in the perfect wave function, as one can see in terms of the configuration interaction wave function when we write

$$\Psi_{\text{exact}} = \Psi_{HF} + \sum_i c_i \Psi_i^{\text{excited}} \qquad (13)$$

and the correction terms on the right-hand side arise[138] by raising one or more electrons into the molecular orbitals which are vacant in Ψ_{HF}. It follows from this that the separation into σ and π-electrons is really artificial and is introduced by the use of a one determinant approximation in the Hartree-Fock method. Thus, one can build up Σ states in which π-molecular orbitals are occupied and Π states with just one π-molecular orbital occupied and all the other electrons in σ-molecular orbitals. So in the final perfect wave function the distinction between σ- and π-electrons has disappeared as the orbital concept has disappeared and so the final concept of aromaticity will not be expressed in terms of either σ- or π-electrons but in terms of electrons.

3.4 The Current Situation

The best that we can do today, in any systematic manner[131, 132], is the closed-shell one-determinant Hartree-Fock approximation. In accepting this picture, we accept the orbital picture and so think of each electron as inhabiting an orbital.

In the self-consistent field method (chapter 3.3) the wave function is represented by a single determinant of doubly occupied molecular orbitals, some of σ-type and some of π-type. From equation 12, it is easy to see that the individual molecular orbitals may (not must) be chosen to satisfy the eigenvalue equation

$$f(1)\,\phi_k(1) = e_k \phi_k(1) \qquad (14)$$

where f(1) is the Hartree-Fock operator for electron numbered one, ϕ_k is the kth molecular orbital and e_k is approximately the ionisation potential of this molecular orbital. The detailed form of the operator is given in equation 12.

Equation (14) is the standard non-empirical equation for the self-consistent field molecular orbitals of a closed-shell molecule. Extensive experience with this equation[137,138] has shown that the molecular orbitals and the total wave function generally account for just over 99 per cent of the total molecular energy. The remaining 1 per cent is the correlation energy plus the error arising from the approximate forms of the orbitals. Introducing the lcao (linear combination of atomic orbitals) approximation leads directly to the usual secular equations and secular determinant.[29]

The solution of this equation for benzene and similar molecules has opened up a new chapter in the computations of the wave functions of aromatic molecules. There is still a large error in absolute terms in the total energy but the entire procedure is well defined so that the removal of the remaining error, difficult though it is at the moment, is a well defined procedure.[137, 138]

The results which emerge from the computation are the forms of the molecular orbitals, the energies of the molecular orbitals and the total energy of the molecule. If other expectation values, $\bar{M}$ say, of the operator M_{op} are required, they are computed from the usual equation[29] (15)

$$\bar{M} = \langle \Psi \; M_{op} \; \Psi \rangle \qquad (15)$$

The forms of the molecular orbitals are themselves quite complicated and it is usually difficult to see just what the significance of the form of a particular wave function is in physical terms. Some workers feel that we ought to be content with computed values for observable quantities such as dipole moments, but the majority of people seem to feel that we should work out the charge or population of an atom together with the bond order of overlap population of a particular bond.[139]

The results for benzene are given in table 7 and those for pyridine and pyrrole in tables 8 and 9. In all three cases, the reported material is not exactly what we would like to have, because the σ-electrons are not localised into two electron chemical bonds, but it is the best currently available.

Table 7. *All-electron, all-integral SCF MO computations on the benzene molecule*[131]

Energy of M.O.s. in a.u.		Populations		
$1e_{1g}$ (π)	-0.375	Carbon atoms	$2p\pi$	= 1.000
$3e_{2g}$ (σ)	-0.525		$2p\sigma$	= 3.23
$1a_{2u}$ (π)	-0.536	Hydrogen atoms	$1s$	= 0.77
$3e_{1u}$ (σ)	-0.624			
$1b_{2u}$ (σ)	-0.657			
$2b_{1u}$ (σ)	-0.668			
$3a_{1g}$ (σ)	-0.740			
$2e_{2g}$ (σ)	-0.847			
$2e_{1u}$ (σ)	-1.040			
$2a_{1g}$ (σ)	-1.170			

1 a.u. = 27.21 eV = 2624.5 kJ mol^{-1}

A major piece of information on the benzene molecule is the energy levels of the molecular orbitals. The significant result here is that, although the highest occupied molecular orbital is indeed a π-orbital, there is a σ-level in between the two π-electron levels. This is a clear warning that the low lying ionised and excited states of the molecule may contain large σ-electron components.

The populations of the atoms show strongly polar $C^{-}H^{+}$ bonds just as in the methane case.[33] We cannot derive the hybridisations from the reported molecular orbitals. This is unfortunate and a localised version of the π-molecular orbitals would indeed give this information providing some precautions were observed.[33] It is generally true that one cannot separate charge transfer and hybridisation in delocalised molecular orbital formulations but these two quantities may be separated easily in the localised formulation.[33]

The pyridine results (table 8) show rather similar effects. There is a closely grouped set of two π-molecular orbitals and one σ-molecular orbital at about 0.45 atomic units (~12 eV). And there is another σ-level between this group of three levels and the third π-electron molecular orbital. So again we have an intimate mixing of the σ- and the π-molecular orbitals on the energy scale. This implies quite clearly that the low lying ionised and excited states may contain much σ-character.

Table 8. *All-electron, all integral SCF MO computations on the pyridine molecule*[132]

Energy of M.O.s. in a.u.			Populations			
a_2 (π)	=	-0.4473	Nitrogen atom	$2p\pi$	=	1.010
b_1 (π)	=	-0.4586		$2p\sigma$	=	2.709
a_1 (σ)	=	-0.4654		$2s$	=	1.509
b_2 (σ)	=	-0.5795	Carbon (1,5)	$2p\pi$	=	1.005
b_1 (π)	=	-0.6223		$2p\sigma$	=	2.058
a_1 (σ)	=	-0.6394		$2s$	=	1.047
b_2 (σ)	=	-0.6700	Hydrogen (1,5)	$1s$	=	0.778
a_1 (σ)	=	-0.7012	Carbon (2,4)	$2p\pi$	=	1.002
b_2 (σ)	=	-0.7260		$2p\sigma$	=	2.158
a_1 (σ)	=	-0.7792		$2s$	=	1.066
b_2 (σ)	=	-0.9044	Hydrogen (2,4)	$1s$	=	0.783
a_1 (σ)	=	0.9218	Carbon (3)	$2p\pi$	=	0.975
b_2 (σ)	=	-1.1103		$2p\sigma$	=	2.155
a_1 (σ)	=	-1.1577		$2s$	=	1.072
a_1 (σ)	=	1.3277	Hydrogen (3)	$1s$	=	0.780

The π-electron populations in pyridine are nearly one π-electron per ring atom as in benzene and this uniformity of π-electron populations is reminiscent of the small molecular results.[140] Evidently the nitrogen atom's electronegativity is satisfied by the σ and not the π-electrons in the same way as in the smaller molecules. No doubt this is why the electronic spectrum of pyridine is very like that of benzene. The carbon-hydrogen bonds are strongly polar in the C^-H^+ sense in pyridine.

The pyrrole results (table 9) show a difference in that there is no high energy

Table 9. *All-electron all integral SCF MO computations on the pyrrole molecule*[132]

Energy of valence shell M.O.s in a.u.			Populations		
a_2 (π)	=	-0.3879	Nitrogen atom	$2p\pi$	1.659
b_1 (π)	=	-0.4253		$2p\sigma$	2.380
a_1 (σ)	=	-0.5766		$2s$	1.371
b_2 (σ)	=	-0.6022	Hydrogen (N)	$1s$	0.661
b_2 (σ)	=	-0.6243	Carbon (1,4)	$2p\pi$	1.075
b_1 (π)	=	-0.6313		$2p\sigma$	1.984
a_1 (σ)	=	-0.6476		$2s$	1.047
a_1 (σ)	=	-0.7779	Hydrogen (1, 4)	$1s$	0.796
b_2 (σ)	=	-0.7970	Carbon (2, 3)	$2p\pi$	1.095
a_1 (σ)	=	-0.8251		$2p\sigma$	2.105
b_2 (σ)	=	-1.0345		$2s$	1.056
a_1 (σ)	=	-1.0955	Hydrogen (2,3)	$1s$	0.808
a_1 (σ)	=	-1.3239			

σ molecular orbital despite the fact that there are three σ levels mixed in with the three π levels. The populations show that the nitrogen atom loses π-electrons to the ring and gains σ-electrons from the ring. The carbon-hydrogen bonds are strongly polar in the sense C^-H^+ again.

A number of other *ab initio* computations on aromatic type molecules have been reported and a review has appeared.[140a] The results for the cyclopentadienyl anion[140b] show the same intermingling of the σ and π orbitals as do the fulvene results.[140c] In the naphthalene and azulene results[140d] there is much less mingling of the σ and π-orbital energy levels and in these two molecules the classical idea of the separation of σ and π levels is a good approximation.

The next step is to develop a quantitative measure of the relative importance of the σ and π-electrons in the molecules. A detailed analysis of the total energy is difficult because of the uncertainty of how to divide up the large σ-π interaction energy (*cf* chapter 3.5). A second major complication is that one does not generally have the σ-electron orbitals in localised form but rather in the canonical delocalised form (but see following paragraph). In this situation, the best one can do is to try to extract information from the *e* values of the molecular orbitals and it is natural to use the sum of the *e* values (Σe) for the various types of molecular orbitals in a molecule.

Table 10. *The total orbital energy (Σe) of the benzene and fulvene orbitals*[140c]

	Benzene	Fulvene	Difference	Difference per orbital
$-\Sigma e^{\pi}$	9.608	9.637	-0.03	-0.01
$-\Sigma e^{CH}$	13.379	13.314	0.065	0.01
$-\Sigma e^{CC}$	30.869	30.497	0.37	0.06
$-\Sigma e^{1s}$	175.844	175.788	0.06	0.01

Atomic units (27.21 eV).

This can be done most clearly when the σ-orbitals are in localised form.[140c] In table 10 we give the comparison between the (Σe) for the two C_6H_6 molecules benzene and fulvene. As far as the π-electrons are concerned, it seems that the quantity (Σe) is about the same in the two molecules and the same is true for the carbon-hydrogen bonds. But the benzene molecule is pronouncedly more stable in the carbon-carbon σ- bonds than is fulvene. The inference then is that the stability of benzene relative to fulvene is in the σ-electrons and not in the π-electron frame. The magnitude of the difference is too large by a factor of three or four although the total energies from the computations do reflect the experimental energy differences well.

These results are only just beginning to throw light on our ideas about aromaticity and extensive development is to be expected in this area shortly. One difficulty with nearly all the results so far is that the σ-orbitals of say benzene and naphthalene appear to be quite different and unrelated to each other. Yet experience says clearly that the σ-electrons in the two molecules are similarly organised. The difficulty here lies in the use of the delocalised or canonical molecular orbitals rather than the localised orbitals and so concealing the chemically important similarities between these two molecules. So once again the use of the localised molecular orbitals[33] is indicated.

3.5 Lower Approximations and $\sigma - \pi$ Separability

The connection between the rigorous theories of sections 3.3 and 3.4 and the semi-empirical theories of sections 3.7, 3.8 and 3.9 is rarely discussed in general texts yet it is the cornerstone on which all the lower theories are built. In particular, it would be very helpful if we could justify the successful looking semi-empirical theories in terms of the non-empirical ones. Broadly speaking, this cannot be done at all thoroughly at the present time and the material of this section is an attempt to familiarise the reader with the problems which arise in this area.

All of our difficulties stem from the fact that it will be a long time, if ever, before we can routinely solve equation 14 in all its generality for any molecule. For practical purposes, we must rely on approximate solutions of this approximate equation. However, as we begin to introduce approximations, we lose touch with the rigorous (but not exact) theories of sections 3.3 and 3.4. This general situation has led to the growth of a bewildering variety of lower approximations many of which seem to be virtually the same but which sometimes give different numerical answers in the hands of different workers. Indeed, some workers do not define their approximations and others seem to change the definitions so that the answers fit with experimental values or preconceived ideas. This is not a happy situation and one particular casualty of this confusion has been the aromaticity concept in general, and, in particular, the concept of resonance energy.

The lower approximations are often characterised by whether the method is or is not an iterative or self-consistent one.[135] If the method is self-consistent, then the conclusions which emerge at the end are consistent with the model which was assumed at the beginning. Thus, the Hückel method is not self-consistent but the Wheland-Mann method or ω-technique is self-consistent to the extent that the π-electron charges are made self-consistent (*cf* section 3.9). But this question of self-consistency is really a technical mathematical one which has little physical significance, at least as far as we know, in the general case. There is another characteristic of the lower theories which has physical significance.

This characteristic is that nearly all of the lower theories presuppose some kind of separation of the 'π-energy' and the 'σ-energy' by writing down, or assuming implicitly, the equation

$$E_{\text{total}} = E\sigma + E\pi \qquad (16)$$

Discussions of this point can be found in many places in the literature.[141] In addition, we must write down an eigenvalue equation such as

$$h\phi_i = e_i \phi_i \tag{17}$$

where h is a one electron operator, ϕ_i is the ith molecular orbital and e_i is the eigenvalue which represents the energy of this molecular orbital.

These two equations, (16) and (17), look quite simple at first sight, being similar to the equations (9) and (14) of the rigorous theory. But they are in reality a fruitful source of confusion, particularly to the someone who is not primarily a theoretician.

Taking the eigenvalue equation (17) first, no difficulty arises with this providing it is supposed that it is identical with the formal equation (14) whose meaning is exactly defined. Then we know just what the operator is, we know that the semi-empirical molecular orbital should be identical with the molecular orbital of the formal theory and we also know that e_i is the ionisation energy of the electron in the ith molecular orbital, at least for the closed shell case. This last result must follow if equations (14) and (17) are to be the same thing. So we should presumably test the accuracy of this result that the eigenvalue in the ionisation energy as our first step in evaluating the performance of a semi-empirical theory. Thus, in Hückel's π-electron theory (*cf* section 3.9), the quantity $(m_i a + n_i\beta)$ where a is the coulomb integral, β is the resonance integral and m_i and n_i are the numbers obtained from the secular determinant, should give the ionisation potential of the ith molecular orbital. Tests of this point may be found in the standard literature.[135]

But the quantity e_i, the eigenvalue, is in practice given a wide variety of interpretations.[135] For example, it is often supposed that this quantity is related to atomisation and bond energies by taking the n_i part as the bond energy. Going on from this, it is also often supposed that the differences which are found in the quantity n_i between two molecules represent differences in the resonance energies of the two molecules. Alternatively, a difference such as $(e_i - e_j)$, where the jth molecular orbital is an empty orbital, is often supposed to represent an excitation energy. In neither case is there a theoretical basis for these conjectures and this point should be appreciated clearly.

To summarise, if we are to keep contact with the rigorous theories, then the e_i must be taken as the ionisation energy of the molecular orbital.

A similar situation arises with the total energy equation. Comparing equations (9) and (16) we see that the difficulty in equating these two equations lies in the disposal of the large $\sigma - \pi$ interaction terms of equation (9) which is written in detail as (*cf* equation 10)

$$2\sum_{\sigma}\sum_{\pi} (2J_{\sigma\pi} - K_{\sigma\pi})$$

This is a very large amount of energy, hundreds of electron volts even in a diatomic molecule and it represents the interaction between all the σ-electrons and all the π-electrons. The task of trying to split up this amount of energy in to a part belonging to the σ-electrons and a part belonging to the π-electrons is a most delicate one,[141] and various attempts have met with complexities and difficulties.

It follows that if equation (**16**) is assumed, it has no theoretical basis and its success is to be judged solely by the conclusions drawn from it being approximately in agreement with experiment.

To see just one example of the physical effects of the $\sigma - \pi$ interaction we can look at the results of the analysis of the accurate small molecule computations. Using as an example the formaldehyde molecule, it is commonly supposed that the π-bond is polarised C^+O^-. This is not born out by the detailed computations[142] of the molecule's wave function, the analysis[143] of which suggests that the σ-bond is so heavily polarised C^+O^- that the electronegativity of the oxygen atom towards the π-electrons is about the same as that of the carbon atom towards the π-electrons. The overall result is that the π-bond of formaldehyde is non-polar. Much the same happens in other small molecules,[144] and the same thing happens in pyridine where the π-electron populations are uniform (table 8). If we turn to the more elaborate computations on such molecules as formaldehyde[145], it is unfortunately difficult to analyse the complicated wave functions in these simple terms.

It is interesting to notice that some semi-empirical methods[146] give much the same result, although here the parameter choices must be the deciding factor in producing this result. Experimental results bearing on this point are available.[147] These examples show clearly just how drastically the σ-electron distribution can affect the π-electron distribution and computations in which the σ-electrons are ignored can give misleading results, particularly when hetero-atoms are involved.

Despite this critcism, it is certainly true that equation (16) describing a simple additivity of E^σ and E^π has been widely used in an empirical spirit. But a related problem now arises of deciding just what E^σ and E^π do represent. The first thought is that, taken together they must represent the total energy of the molecule (the energy to separate all the electrons and nuclei) as defined in equation (**9**). If this is so, then both quantities must represent huge amounts of energy – hundreds of electron-volts (eV) even in small molecules. But such quantities of energy are not those commonly used in chemistry. These latter amount to several eV per bond. Since we have sunk deep into empiricism by using equation (**16**), might it not be just as sensible to suppose that E^σ and E^π really represent bond energies? The legitimacy of this procedure is impossible to assess.

To summarise it seems impossible to retain detailed connections with the formal theory while using the semi-empirical methods.

We have stressed how important is the role of $\sigma - \pi$ interaction in the rigorous theory. Following on this point of view, we have divided up the remaining approximation methods according to how well they treat this $\sigma - \pi$ division. The crudest method is that of ignoring the σ-electrons altogether, as in Hückel's method of section 3.9. An intermediate level of approximation is provided by the theories in which the σ and π-bonds are both supposed to have 'natural' bond lengths and the real bond length is a compromise between the two. These are given in section 3.8. The most sophisticated theories are those in which the σ and π-electrons affect each other in detail but still without the full use of the rigorous equation (**9**) as in section 3.7.

Before completing the work with these three sections we can now discuss the entire question of resonance energy and stabilisation energy.

3.6 Resonance and Stabilisation Energies

Now that we have a clear picture of the connection, or lack of connection, between the various sorts of 'energy' which arise in the lower theories, it is possible to discuss resonance and stabilisation energies. This is perhaps one of the most confused subjects in chemistry, yet the essentials were set out clearly some years ago.[148] The difficulties arise partly from mixing together empirical ideas and theoretical methods without the boundary between the two being clear, and also from the well-established habit of mixing together valence bond and molecular orbital concepts.

This material is a continuation of that of chapter 2 where we pointed out that it is difficult to get accurate reliable values for the total energy of hypothetical molecules such as cyclohexatriene. The basic idea of resonance energy is derived by considering two molecules (**8**) and (**19**), (*cf* figure 12 of chapter 2.3.1), and defining the difference between the total energy of these two molecules as the resonance energy of benzene. Note that the reference molecule is not cyclohexatriene with its equilibrium bond lengths. Notice also that the drawing of pictures such as those of figure 12 is not an *exact* representation of the way in which the electrons are organised so that neither molecule is exactly defined by these pictures. Representing all three molecules on an energy scale gives figure 29. The rigorous definitions of stabilisation energy and of resonance energy are given in this figure.

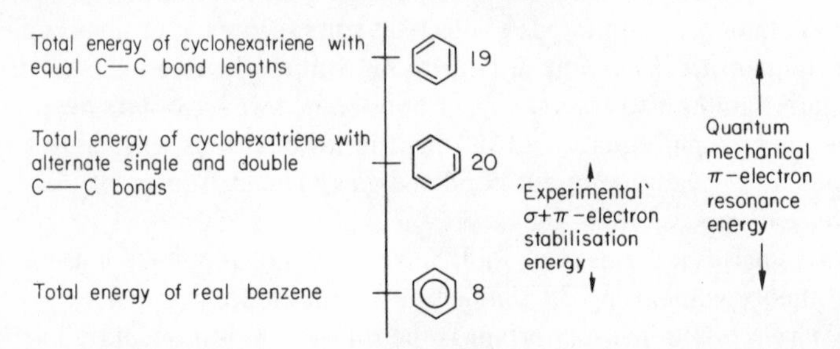

Figure 29. Qualitative energy level scheme relating to the resonance energy of benzene

In chapter 2, we showed how to perform the resonance energy calculation which gives the classical value of 36 kcal mol^{-1} for benzene. This quantity is better called the stabilisation energy and is represented by the difference in total energy between (**8**) and (**20**). In deriving this number we ignored the σ-electrons entirely and we have just seen that this omission has some serious consequences, since the quantum mechanical resonance energy is estimated[148] at 63 kcal mol^{-1}. This is the difference in total energy between (**8**) and (**19**).

Other doubtful points arise in trying to work out the energetics of hypothetical molecules. One point which limited experience suggests is important is the type of hyperconjugation energy now known as sigmaconjugation energy.[72] This arises because no bond is ever completely localised and it seems that this effect may well be quite large.[33] To say what is much the same thing in another way, when we break a carbon-hydrogen bond in ethylene, there is significant reorganisation of the other electrons in the adjacent bonds. When we then put the fragments together again, as in figure 13, there may be further and partly compensating rearrangements of the σ-electrons. The size of these effects in benzene is completely unknown.

Another point is the question of non-bonded interactions which are of considerable importance in conformational studies[149] but are usually neglected in resonance energy studies (steric inhibition of resonance apart). The dipole-dipole interaction terms (and perhaps higher multipole terms) are neglected and, although these seem to be small in general,[33] we have no idea how large they are in benzene. Finally, the extent of the correlation energy changes in going from ethylene to benzene are unknown. It is admittedly true that some of these particular difficulties might be avoided by using a more elaborate reference standard than ethylene, such as cyclohexene or linear hexatriene, but this only moves the difficulties to the question of the nature of these reference molecules. In general, it seems best to have the simplest possible comparison molecules.

Another difficult point concerning resonance energies is the question of whether they should be given as per ring, with or without regard to ring size, or per sextet (cyclic C_3^+?) or per π-electron. This is an awkward point which lacks a convincing answer as far as we know.[150]

An important point about resonance energies which is rarely appreciated is that one could have a large resonance energy (or stabilisation energy or delocalisation energy) in a molecule which is not aromatic in our sense of that term. An example of this and some other points is the interesting work[150a] on the guanidinium cation (**36**). This is stable molecule and it has 6π-electrons in four parallel $2p\pi$ atomic orbitals. The exact form of the molecular orbital diagram will depend on details of the computation but the general appearance of the molecular orbital diagram is largely determined by symmetry and must be as given in figure 30. This diagram looks like that of benzene to some extent so one might be inclined to call this and similar molecules[150b] 'aromatic' or 'aromatic from the centre'.

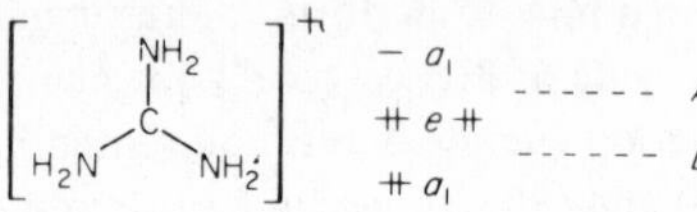

Figure 30. Vertical axis: orbital energy (C_{3v} not D_{3h} classification)

There is a difficulty here in that much depends where one puts the zero line onto the diagram of figure 30. If one puts it high up as on the line labelled (**A**), then the resemblance to the benzene diagram is marked but if one puts it down low in the position marked (**B**) then all resemblance to the benzene diagram is lost. Since the one electron energy of semi-empirical molecular orbital theory is an ill defined quantity, one cannot be certain where to put this zero line particularly when heteroatoms are present as in this case. An equally difficult point is to know what to use as a reference molecule for delocalisation energies.

This particular case illustrates how subtle and deceptive semi-empirical molecular orbital theories can be and it is clear that this molecule may well not be 'aromatic' in any sense.

3.7 CNDO Theory

The underlying idea of this theory was developed originally[151] for the computation of the excited states of π-electron systems. It has now been extended for general use in calculating molecular wave functions.[152] The basic idea is to get as close as possible to the formal equations of section 3.3. without getting too involved in the integral evaluation problems of that theory. To do this, one neglects all integrals over atomic orbitals which contain in the integrand an overlap charge. For example, in the π-electron case an integral such as $\langle\pi_a\pi_b/\pi_b\pi_b\rangle$ which describes the repulsion between the charge cloud ($\pi_b{}^2$) on atom b and the overlap charge cloud ($\pi_a\pi_b$) is put equal to zero. This approximation is made systematically throughout all the integrals which occur in the theory and it is hoped that this consistency will partly compensate for the error within the theory. Detailed descriptions of the method and of its performance are available in the literature.[153]

There are as yet rather few applications of this method to all the valence electrons of aromatic systems[154] and even then resonance energies are often not reported. Examples are given in table 11.

Table 11. Examples of resonance energies from various sources

Molecule	Simple Hückel (in units of β) Ref.27 p.226; Ref.29 Chap.3	Ref.158 (in eV)	Ref. 141b (Tables X & XI) (in eV). 'Hückel'	'CNDO'	Ref. 141c (Tables IX & XI) (in eV) 'Hückel'	'CNDO'	Ref.11 p.194 (in eV)	Ref.141d (Table 2) (in eV)
benzene	2.00	2.05	1.32	1.32	1.32	1.32	0.97	0.87
naphthalene	3.68	3.69	2.31	2.43	2.41	2.28	1.46	1.32
anthracene	5.31	5.17	3.26	3.38	3.78	3.09	1.78	1.60
phenanthrene	5.45		3.37	3.75	3.89	3.45	2.13	1.93
azulene	3.36		2.06	1.36	2.17	1.22	0.30	0.17
fulvene	1.47		0.89	0.14	0.90	0.14		0.05
pentalene	2.46		1.43	0.73	1.53	0.59		0.01
$C_5H_5^-$	2.47							
$C_7H_7^+$	2.99							

$\beta \sim 20$ kcal mol^{-1}

3.8 Theories with σ-Electrons Treated as Simple Two-Electron Chemical Bonds

These theories[154a,b,] are those in which we do not attempt to solve the rigorous equation (14) even in an approximate form, but we accept as starting point the equation (16).

$$E_{\text{total}} = E^{\sigma} + E^{\pi}$$

We can now freely interpret E^{σ} and E^{π}. Many different procedures can arise here but the most common is to take E^{σ} as the 'σ-bond energy' and E^{π} as some calculated 'π-bond energy'. As was stressed earlier, we have now lost all contact with the rigorous theory and we retain only its mathematical apparatus.

The simplest examples[155] of this theory begin by assuming that the σ-bonds have a 'natural' length, r^{eq}, a 'natural' bond energy and a 'natural' force constant, k.

$$E^{\sigma} = \sum_{p}^{\text{bonds}} k_p (r_p^{eq} - r_p)^2$$

where r_p is the actual length of the pth bond in the real molecule.

This general method was first used in 1937[155] and has been extended considerably in various ways.[156] One interesting result which is relevant to aromaticity is that a cyclic polyene with $(4n+2)$ atoms has alternating double and single bonds in the large members with n greater than about six. These conclusions are sensitive to the assumptions which are made in this theory and they must be accepted with reserve.

A more sophisticated version of this type of theory[157] is based on the idea that the carbon-carbon σ bonds are different in benzene from those in ethane, and acetylene. If we accept the idea that the carbon atom uses sp^3, sp^2 and sp hybrid atomic orbitals to form its bonds in these molecules[157a] then we should presumably try to assign a 'natural' length, a 'natural' bond energy and a 'natural' force constant or constants to each of the sp^3-sp^3, sp^3-sp^2, sp^3-sp, sp^2-sp, sp-sp carbon-carbon bonds which will arise in the general case and similarly for the three different types of carbon-hydrogen bonds. We have altogether eight different types of bonds, with at least three parameters to assign for each, a minimum of twenty four parameters. This large number is reduced by making further assumptions about the parameters for different bonds being the same until we have a workable number of parameters, say a dozen, which can be fixed by comparison with some simple molecules such as ethane, ethylene, butadiene, etc. The recommended values of the parameters for some bonds are given in reference 141 and 157 (*cf* table 11). Using these recommended values, computations on large molecules can be made in a straightforward way (table 11).

The complexities and uncertainties of these procedures seem to have led many workers to feel disillusioned with the whole idea of treating the σ-framework of π-electron systems as simple two-electron chemical bonds and many workers have turned to the all-valence electron semi-empirical methods in which all the valence electrons are delocalised. This is perhaps unfortunate because we now have an accurate theory of chemical bonds[33] which we may use in the π-electron context, at least for the σ-electrons, and so keep the whole theory as chemical as possible.

This general type of theory has been used very widely and there are many examples in the literature although again in many instances the problem of resonance energy is not discussed. Some relevant results are given in table 11.

3.9 Theories in which the σ-Electrons are Neglected

These are the oldest and simplest of all the many theoretical methods which have been used to calculate the wave functions of the large π-electron systems.[25] Again, there are many possible variants within this general theory.[135]

From what has been said so far, it may seem that the use of such theories as these is an act of intellectual heroism and that the results are of historical interest only. But matters are more complicated than this because these simple methods have one major advantage over and above the practical consideration that they are very easy to use.

This advantage is that they contain very few parameters whose numerical values can be adjusted until agreement with experiment is reached. This is an important consideration because, although it is difficult to express the idea quantitatively, most experienced workers seem to feel that if you have enough disposable parameters the theory can be made to yield almost any numerical results that one wishes and in particular, one can get agreement with almost any experimental facts. A clear example of this situation is found in the literature of the 1930s where calculations of such quantities as activation energies are reported and these calculations give agreement with experiment to within a few kcal mol^{-1}. Today, these computations are not taken seriously. It is certainly true, however, that the advantage gained by reducing the number of parameters helps to make up for the loss of rigour in these theories.

The σ-electrons being ignored, the π-electrons can then be treated at various levels of approximation. The original and simplest of these methods is Hückel's[25,159], the details of which are given in many texts, and which led to the (4*n*+2) rule of aromaticity and the energy level diagram for the C_5^-, C_6 and C_7^+ cycles shown in figure 31.

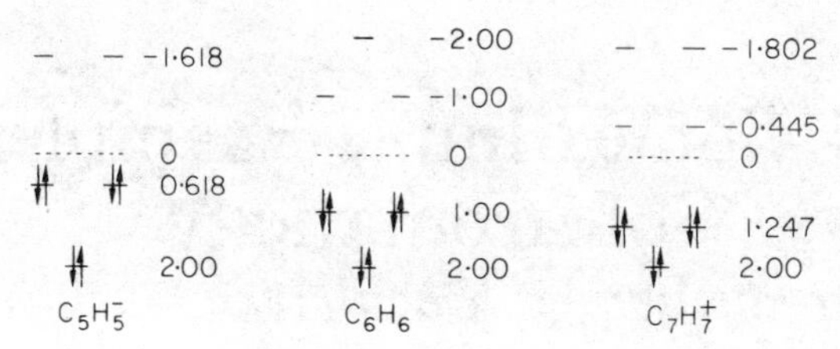

Figure 31. Vertical axes: orbital energies in units of β

The results relevant to aromaticity which follow from Hückel's theory are also well known[135] and a few typical examples are reproduced in table 11.

A refinement is the inclusion of some charge self-consistency in the π-electrons and this is the Wheland-Mann[160] or ω-technique which also has its variants [161].

A further refinement is the Coppert-Mayer and Sklar work[162] on benzene in which the π-electrons are treated at the full self-consistent field level of approximation but this is more relevant to excited state work than to ground states and to aromaticity in general.

3.10 Summary

In this section, we have laid stress on the difficulties and uncertainties which even the best available theoretical methods meet with in molecules as large as the aromatics, where disagreement between similar theories is commonplace.[163] We do this, not from a sense of pessimism about the future of the general topic of aromaticity but rather to counterbalance what seems a general atmosphere of credulity and naivete in the use of the semi-empirical theories. In the long term, this situation can lead only to widespread disillusion with the entire subject of wave function calculation and perhaps to the falling into disrepute of this important topic in chemistry

4. Some Simple Examples of Aromaticity

Introduction – phenanthrene – azulene – [18] annulene – ferrocene – thiophen – fulvene – pentalene – summary

4.1 Introduction

This section is concerned with the comparison between experimental facts and theoretical ideas. We have chosen a few examples which illustrate the various points as directly as possible.

In effect, we have put ourselves in the position of someone who is confronted with a new molecule and wishes to know just what experimental facts to gather and what theoretical ideas to consider in trying to decide whether it is aromatic.

The molecules shown in the heading are the main topics of each subsection but closely related examples are sometimes introduced.

4.2 Phenanthrene (37)

This is an example of a classical aromatic molecule which satisfies the $(4n+2)$ rule as this is widely used. Strictly speaking, the original derivation of the $(4n+2)$ rule applied only to monocycles but it is often extended,[164,165] to the cross-linked molecules such as phenanthrene.

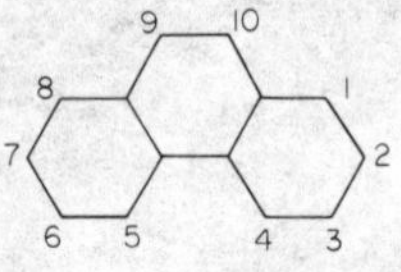

9 10 8 1 7 2 6 5 4 3

37

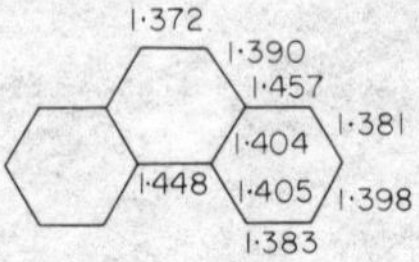

1·372 1·390 1·457 1·381 1·404 1·448 1·405 1·398 1·383

Figure 32.

Table 12. *Thermochemical data on the examples of chapters 4 & 5.**

Molecule	ΔH^{o}_{comb} (c)	ΔH^{o}_{sub}	ΔH^{o}_{comb} (g)	Calc ΔH^{o}_{comb} (g) Klages/ Franklin (ref.70)	Resonance Energy Klages Franklin
2-aminotropone	889.9[a]	17.0[a]	906.9	923.5/926.6	16.6/19.7
azulene	1265.1[b]	18.4[b]	1283.5	1312/1316	28.5/32.5[≠]
p-benzoquinone	656.3[c]	15.0[d]	671.3	676/675	4.7/3.7
tropylidone	965.6[a]	9.25[e]	974.85	981.4/983.3	6.6/8.5
2, 7-dimethyl-4,5-benzotropone	1613.3[f]	18.7[f](est)	1632.0	1681.8/1680.3	49.8/48.3
6, 6-dimethyl-fulvene	1116.1[b](1)	10.6[b](vap)	1126.7	1135.9/1141.8	9.2[I]/15.1[I]
ferrocene	1404.8[g]	17.4[h]	1422.6	–	–
phenanthrene	1686.1[b]	21.7[b]	1707.8	1797.7/1800.5	89.9/92.7
porphin	2094.7[i]	26.0[j]	2120.7	2625.0/***	504.3/***
thiophen	676.1[k](1)	8.30[l](vap)	684.4	640.7/638.9	27.2**/25.4**
tetracyano-ethylene	713.4[m]	19.4[m]	732.8	699.6/707.1	-33.2/-25.7
tetracyanoquino-dimethane	1424.2[m]	25.1[m]	1449.3	1460.6/1469.9	11.3/20.6
tropolone	808.3[a]	20.1[a]	828.4	846.8/847.8	18.4/19.4
tropone	860.9[a]	12.9[a]	873.8	886.9/888.4	13.1/14.6

* All values in kcal mol^{-1}. The details of the symbols are explained in the footnote to table 2.

** For sulphur containing compounds, Wheland's tables[70] apply when SO_2 (g) is a product rather than H_2SO_4 (aq). The observed $-\Delta H^{o}_{comb}$ (g) applies to the latter process. For reaction $SO_2 (g) + (1/2)O_2 (g) + H_2O + aq \rightarrow H_2SO_4$ (aq), $\Delta H = -73.2$ kcal mol^{-1}, but Wheland uses a value of -70.9 kcal mol^{-1}, and this latter value has been used here.

I Heat of hydrogenation and other data suggest a value of 11.5 kcal mol^{-1} (ref. 69, p.72).

≠ Compare with value from heat of hydrogenation (table 4).

*** Franklin's Scheme not applicable

References for Table 12

a. Jackson W., Hung T.S. and Hopkins, Jr., H.P. (1971) *J. Chem. Thermodynamics,* **3**, 347.

b. References in table 2.

c. Pilcher, G. and Sutton, L.E. (1956) *J. Chem. Soc.* 2695

d. Magnus, A. (1956) *Z. Phys. Chem. (Frankfurt)* **9**, 141

e. Finke, H.L., Scott, D.W., Gross, M.E., Messerly, J.F., Waddington, G. (1956) *J. Amer. Chem. Soc.,* **78**, 5469

f. Schmid, R.W., Kloster-Jensen, E., Kováts, E. and Heilbronner, E. (1956) *Helv. Chim. Acta* **39**, 806

g. Cotton, F.A. and Wilkinson, G. (1952) *J. Amer. Chem. Soc.,* **74**, 5764

h. Andrews, J.T.S. and Westrum, Jr., E.F. (1969) *J. Organometallic Chem.* **17**, 349

i. Longo, F.R., Finarelli, J.D., and Schmalzbach, E. (1970) *J. Phys. Chem.,* **74**, 3296

j. Edwards, L., Dolphin, D.H., Gouterman, M, and Adler, A.D. (1971) *J. Mol. Spec.* **38**, 16

k. Sunner, S. (1963) *Acta Chem. Scand.* **17,** 728

l. Waddington, G., Knowlton, J.W., Scott, D.W. Oliver, G.D., Todd, S.S., Hubbard, W.N., Smith, J.C. and Huffman, H.G. (1949) *J. Amer. Chem. Soc.,* **71,** 797

m. Boyd, R.H. (1963) *J. Chem. Phys.* **38,** 2529

Phenanthrene is planar and the bond lengths (X-ray) are shown[166] in figure 32. The bond lengths of such molecules as phenanthrene are quite well reproduced both by the simplest Hückel calculations[167] and by the more elaborate computations.[168] A simple comparison with the bond length of benzene (1.395 Å) suggests that the phenanthrene molecule is indeed aromatic as judged by its bond lengths and its planarity.

The resonance energy (table 12) is estimated at 90 to 93 kcal mol^{-1} using the same theory as that which gives benzene a resonance energy of 36 kcal mol^{-1} The use of the more elaborate theories (table 11) gives values of some 80 kcal mol^{-1} for the resonance energy of phenanthrene. We have found no value for the heat of hydrogenation of phenanthrene.

The conclusion is that phenanthrene is aromatic as judged by its resonance energy.

The n.m.r. spectrum or phenanthrene gives the coupling constants[169] and chemical shifts[99] shown in figure 33. Both sets of data support the idea that the molecule is aromatic (*cf* chapter 2.4.1).

^3J values (Hz) τ values

Figure 33. N.m.r. data on phenanthrene

The diamagnetic exaltation, Λ, is 47.3 (table 15) when benzene is 13.7. The diamagnetic anisotropy, ΔK, 165, (table 15) of phenanthrene[170] is about three times the size of that of benzene. Again, both results support the idea that the molecule is aromatic.

The chemical reactivity of the molecule is largely, but not completely, that of an aromatic system. Thus, phenanthrene undergoes[171] acylation under Friedel-Crafts conditions to give the percentage distribution of monoacetyl derivatives shown in figure 34. There are some signs of non-aromatic behaviour in the addition reactions of the 9, 10 region of phenanthrene. Thus, bromination by molecular bromine gives[172] both addition products in the 9,10 position and substitution products with a complex mechanistic scheme.

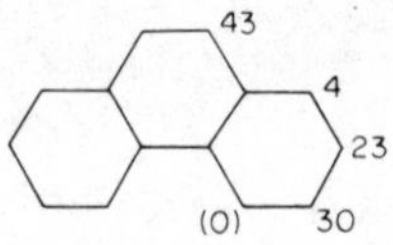

Figure 34. Yield per cent of isomers from monacetylation of phenanthrene[171]

It is not difficult to decide, on the basis of the above evidence, that phenanthrene is indeed an aromatic molecule. The only contrary evidence comes from the chemical reactivity criterion which we feel is perhaps the least reliable criteria of aromaticity.

4.3 Azulene (30)

This molecule[173] is a classical example of a non-benzenoid aromatic which satisfies the ($4n+2$) rule if we disregard the 9, 10 transannular bond. This neglect of the transannular bond is more nearly justified here than in phenanthrene since it is a very long bond, close to the single bond value in length. The bond lengths[174] of the planar substituted azulene **(38)** are shown in figure 35. The fact that chemically equivalent bonds have different lengths is presumably a crystal effect but the size

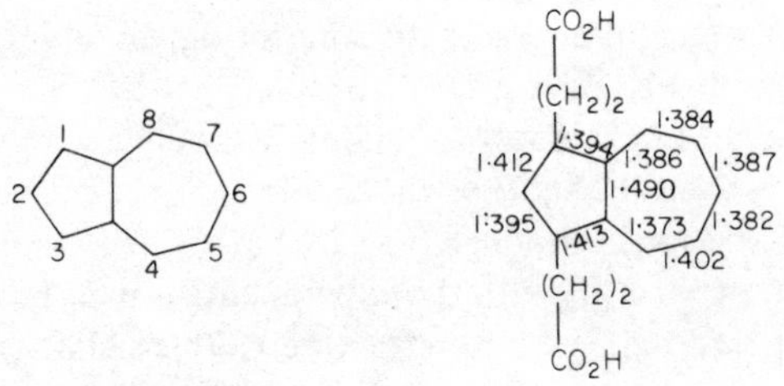

Figure 35.

of this effect is a warning that the third decimal place is meaningless here. The experimental bond lengths are quite well reproduced by simple Hückel calculations[175], using the idea of bond order. More elaborate computations[176] give the same result. A simple comparison of the experimental bond lengths with the benzene value of 1.395 Å strongly suggests that the molecule is aromatic. The central 9, 10 bond is clearly not part of the conjugated system which presumably resembles *[10]* annulene.

The resonance energy of azulene is 30±2 kcal mol^{-1} by the method which gives naphthalene as 61 kcal mole^{-1} (*cf* table 12). An older value[177] of 46 kcal mol^{-1} has been replaced by the new value in the light of recent experimental data. More elaborate computations give[178] the resonance energy of azulene as 7 kcal mol^{-1} while naphthalene is 34 kcal mol^{-1}. Both methods suggest that azulene is considerably less aromatic than is naphthalene.

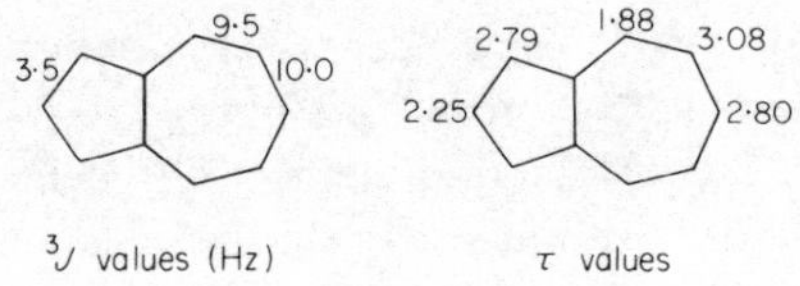

Figure 36.

The n.m.r. data on azulene[179-182] are reproduced in figure 36. The chemical shifts are clearly indicative of an aromatic molecule although it is said that[180] the ring current concept is not adequate in explaining these chemical shifts. Moreover, dilution affects the n.m.r. data significantly in the azulene case. The ^3J coupling constants are those characteristic of a five-membered aromatic ring and a seven-membered aromatic ring. The exaltation, Λ, of the magnetic susceptibility[104] is 26.5, (table 15), and this is clearly in the aromatic category. The anisotropy does not seem to have been recorded. On the whole, the magnetic evidence is strongly in favour of the azulene molecule being an aromatic system.

The chemical reactivity of the azulene molecule is generally felt[183] to correspond to that of a reactive aromatic molecule. Electrophilic substitution occurs readily at the 1(3) position and nucleophilic substitution at the 4(8) position. Addition reactions do not seem to occur in this molecule. The position of substitution agrees well with the calculated atomic charges in the molecule.[184] The chemical reactivity criterion agrees with the other evidence in showing that azulene is aromatic.

The dipole moment of azulene[185] is about 1.0 *D* units with the five-membered ring negative. The value of the dipole moment calculated by Hückel's method[186] is much too high ($\sim 6\,D$) but refined methods[186] give good agreement ($\sim 1.3\,D$) with the experimental value. In valence bond language, it is supposed that the ionic structure (**39**) is a contributor to the ground state. The π-electron dipole moment of such a structure is some 10 *D* units but this would be reduced greatly by reverse σ-electron polarisation so the idea that **39** does contribute significantly to the ground state is a reasonable one.

The azulene molecule is perhaps one of the best possible cases for finding aromaticity in the electronic excited state. It is possible that one of the low lying azulene excited states does correspond more or less to one of the two polar structures (**39**) or (**40**). The structure (**39**) might be expected to be the lower lying of the two since both rings are aromatic but there is some experimental and theoretical evidence in

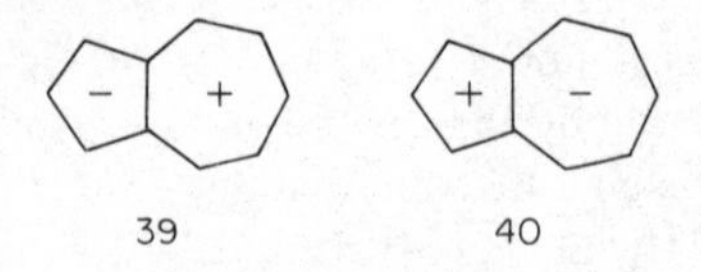

Figure 37.

favour of structure (**40**)[187] If (**39**) is indeed an approximate wave function of the excited state, then this structure will presumably be more aromatic than in the ground state. An experimental measurement of the dipole moment of such an excited state would be most interesting since calculations[188] of this moment suggests a value of some 3 *D*.

Various heterocyclic analogues of azulene have been reviewed.[189] The thio analogue (**41**) is interesting since its chemistry is like that of azulene, the molecule undergoing substitution rather than addition.[190] The electronic spectrum of (**41**) is sensitive to solvents so there may be substantial contribution from the ionic structure (**42**) to the excited state or states. As with azulene itself, it is not clear whether (**42**) or its reversed charge analogue is the observed excited state because there is some evidence[190] in favour of the latter possibility.

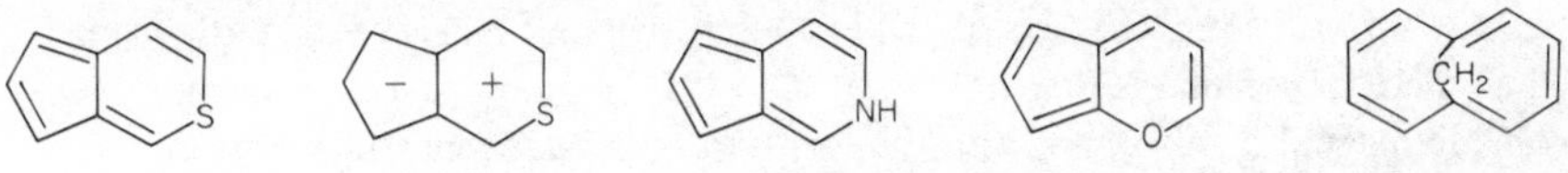

41 42 43 44 45

Figure 38.

The aza analogue (**43**) has been reported[191] and the oxo analogue (**44**) has been subject of computations.[192]

Another interesting relative of azulene is the very unstable [*10*] annulene[193] and such related molecules as (**45**) and those in which the bridging group is oxygen or imino[193a]. The methano derivative (**45**) shows aromatic properties[193a] and the 2-carboxylic acid of (**45**) has an almost planar periphery[194] with bond lengths in the 1.38 to 1.43 Å range. This molecule seems to us to be one of the best pieces of direct evidence that the (4*n*+2) rule has any meaning because it is the only case in which the situation is not complicated by cross bonds or by H-H interactions leading to different conformers.

To summarise the results for azulene, it is clear that this molecule is aromatic on all definitions, at least in a qualitative sense, and this aromaticity seems to occur also in its heterocyclic analogues.

4.4 [*18*] Annulene (26)

This is the best investigated of the large annulenes. It satisfies the (4*n*+2) rule. A detailed review is available.[18] The molecule is nearly planar in the crystal and some of the details of the geometry[18] are shown in figure 39. It is interesting that the *cis* bonds are long

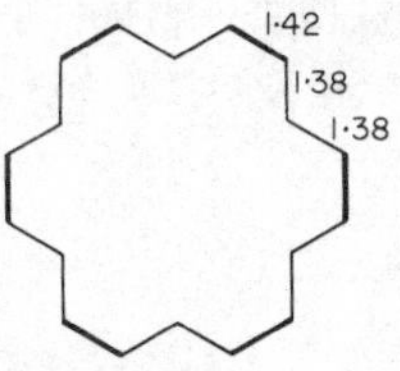

1·42 1·38 1·38

26

Figure 39. Average bond lengths (in Å) for [18] annulene.
The six *cis*-type carbon-carbon bonds are shown by thick lines.

while the *trans* bonds are short and this may simply arise in the σ skeleton where the sigmaconjugation favours the *trans* rather than the *cis* arrangement.[33] The bond lengths strongly suggest that the molecule is aromatic. Computations agree with the observed lengths.[195]

The resonance energy[196] of [*18*] annulene is estimated from experimental data to be about 100 kcal mol^{-1} (table 2). The Hückel method agrees well with the result[197] but more recent computations do not agree with the experimental facts giving a resonance energy of close to zero.[198]

The n.m.r. chemical shifts of [*18*] annulene (table 5) lead to the conclusion that there is a diamagnetic ring current and that the molecule is aromatic.[59] The diamagnetic anisotropy and the exaltation do not seem to have been reported.

The chemical properties depend on the reaction conditions. Under carefully controlled conditions, substitution reactions occur and the mononitration[199a], the monoacetylation[199a], monoformylation[199b] and the monobromination[199b] have all been realised. Under more vigorous conditions, addition products are obtained with bromine[199c] and maleic anhydride gives a Diels-Alder adduct.[199c]

It should be noted[200] that there is a very long wavelength electronic transition at about 13 000 cm^{-1} (λ = 768 nm, ϵ = 14 $m^2 mol^{-1}$). The transition energy is about 1.6 eV or 37 kcal mol^{-1} so this is an extraordinarily low-lying singlet excited state. It seems best to reserve judgement on the nature of the electronic spectra of the large annulenes until it is clear whether the other annulenes have such low lying states. The existence of such states will affect the performance of perturbation theories quite drastically and it must be established whether or not these states exist.

The general conclusion for [*18*] annulene is that the molecule has aromatic bond lengths together with a high resonance energy and a ring current but some of the chemistry is that of a polyene. This seems to be a case in which the chemical reactivity criterion is in conflict with the other criteria and the general conclusion is that the molecule is aromatic. In [*16*] annulene, by contrast, there is almost perfect bond length alternation.[201]

4.5 Ferrocene (46)

This molecule was the first sandwich compound to be synthesised and there is now a vast range of such compounds.[202] The carbon-carbon bond length[203] of 1.440Å is rather larger than that of benzene (1.395 Å) but

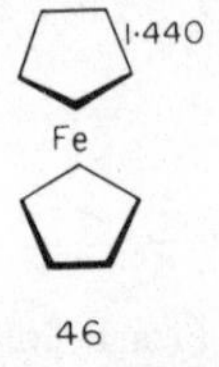

Figure 40.

the C_5 rings are planar regular pentagons. The former result may imply that the rings are not aromatic but the latter result implies that they are aromatic.

The energetics of the ferrocene molecule are not helpful for our present purposes. The heat of combustion[204] in the gas phase is known (table 12) but there is no apparent comparison molecule with which to calculate a resonance energy.

The proton n.m.r. spectrum of ferrocene[205] is a single line at $\tau = 5.94$. It is difficult to be sure what this result means in a molecule as complex as ferrocene.

The diamagnetic exaltation, Λ is not known since no comparison molecule is available with which to compare the known (table 15) diamagnetic susceptibility of ferrocene. The diamagnetic anisotropy has been reported (table 15) giving $\Delta K = 56.0$. Comparing this with the value for benzene ($\Delta K = 58$) suggests a strong ring current in ferrocene.

The chemistry of ferrocene is certainly that of an aromatic molecule.[206] The only qualification which is required is that if the iron atom is oxidised to the ferric condition, the compound becomes altogether more reactive. But ferrocene itself is remarkably 'stable' and 'aromatic'. It is interesting that the chemical reactivity criterion, which we have held in low esteem, is the most straightforwardly informative of all the criteria.

The infrared spectrum of ferrocene is particularly simple[207] but this merely reflects the known high symmetry of the molecule.

We do not yet have a good wave function computation on a molecule as complex as ferrocene but much theoretical discussion and semi-empirical computation has been reported[208]. A difficult point to understand about this molecule is just how the cyclopentadienyl anion preserves its aromatic sextet and so remains negatively charged while at the same time forming strong bonds to iron and not being very reactive towards electrophiles. Perhaps the σ-electrons of the cyclopentadienyl rings confer aromaticity while the π electrons form the bonds to iron (*cf* chapter 3.5).

The general conclusion on ferrocene is that it has some of the properties of an aromatic system but that several of the criteria we ordinarily use to detect aromaticity are not applicable here.

4.6 Thiophen (47)

The geometry of this planar molecule[209] is shown in figure 41.

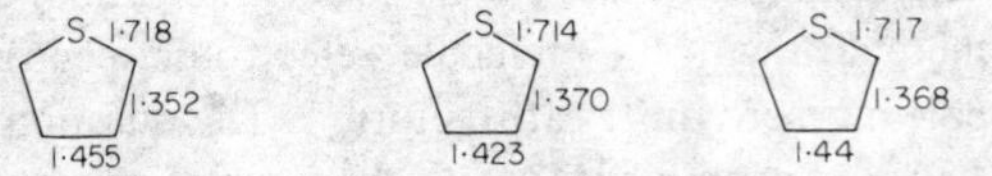

S 1·718
1·352
1·455
47
(Microwave)

S 1·714
1·370
1·423
47
(Microwave)

S 1·717
1·368
1·44
47
(Electron diffraction)

Figure 41.

The different experimental results agree reasonably well among themselves and there is considerable evidence for alternation of the carbon-carbon bond lengths. The 'single' bond is shortened by about 0.03 Å by comparison with butadiene (chapter 2.2). There is certainly no equalisation of the bond lengths as in the classical aromatic molecules.

The energetics of thiophen (table 12) lead to a resonance energy of 36 kcal mol^{-1} using the classical methods of calculation which give a benzene resonance energy of 36 kcal mol^{-1}. On this basis, thiophen is aromatic.

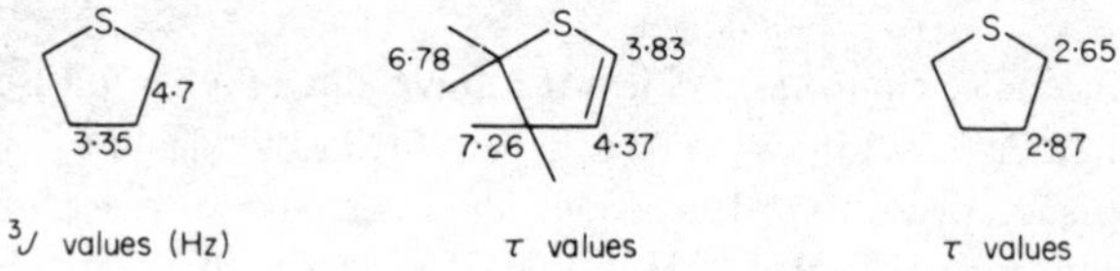

S 4·7 3·35 ^3J values (Hz)

S 6·78 3·83 7·26 4·37 τ values

S 2·65 2·87 τ values

49

Figure 42.

The n.m.r data shown in figure 42 can also be interpreted as those of an aromatic system[210, 211] (*cf* table 5). The diamagnetic anisotropy, ΔK, of thiophen is 37.2 or 50.1 (table 15) (where benzene is 58.4), and the diamagnetic exaltation, Λ, is 13.1.

The e.s.r. spectra of the derivatives of the thiophen radical anion[212] are interesting because they seem to show that the *d*-orbitals do not participate significantly in the wave function of the ground state of the anion. This is an example of the use of e.s.r. spectra to provide information about the electron organisation of a ground state.

The chemistry of thiophen[213, 213a] is certainly that of an aromatic system. The molecule undergoes electrophilic substitution (halogenation, Friedel-Crafts reaction, trifluoroacetylation etc.) rather than addition reaction. Its analogue selenophen has a similar chemistry.[213a]

The dipole moment of thiophen[214] is 0.5 *D* orientated as shown in figure 43. The dipole moment[215] of dialkylsulphides is about 1.5 *D*, presumably orientated as in the figure 43, and the difference between these simple sulphides and thiophen lies in π-electron release into the ring.

S 0·5 *D* 47

R S R 1·5 *D* 50

Figure 43.

Theoretical computations on the π-electron system of thiophen have been reported[214] and these deal with the molecule's electronic spectrum, its bond lengths and its dipole moment quite satisfactorily. The question of the sulphur atom's *d*-orbitals has been examined[216] with the interesting conclusion that the actual contributions of the *d*-orbitals in the molecular orbitals are quite small, but there is a pronounced effect on the calculated values of the electronic properties of the molecule as a result of the inclusion of the *d*-orbitals.

4.7 Fulvene (32)

This molecule is a clear contrast to the first five examples in that it is clearly not aromatic. There is a recent review on the fulvenes[217].

The bond lengths of the dimethyl derivative of fulvene[52] are shown in figure 44. The molecule is planar and the bond lengths are clearly those of a polyene (*cf* figure 7) and not those of an aromatic molecule.

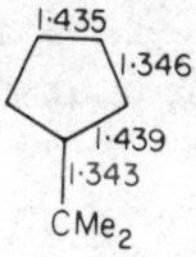

Figure 44.

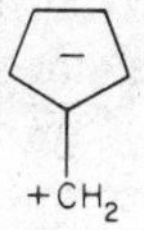

Figure 45.

The calculated bond lengths[218] agree with the experimental ones.

The 'experimental' value of the resonance energy of the dimethylfulvene (**13**) depends to some extent on how it is calculated and values of 9 and 15 kcal mol^{-1} are given by the classical methods (table 12). Another method gives 11.5 kcal mol^{-1} (table 12). These numbers may suggest a small aromaticity but when one realises that resonance energies commonly vary by at least ± 5 kcal mol^{-1} when evaluated by different procedures it is clear that no great significance can be attached to this result. Moreover, a recent calculation[219] on fulvene gives a resonance energy of 1 kcal mol^{-1}.

The n.m.r. spectrum of fulvene[220] may show slight signs of a diamagnetic ring current because the ring protons resonate at $\tau = 3.56$ and $\tau = 3.89$ as compared with the olefinic protons of cyclopentene $\tau = 4.4$ (table 5). The ^3J coupling constants of fulvene (**32**) do not seem to have been reported. Those of 6,6-dimethylfulvene (**13**) are given in figure 25. The static magnetic evidence on fulvene implies a complete absence of aromaticity. Thus, the diamagnetic exaltation, 1.0, is effectively zero (table 15).

The chemistry of fulvene is undoubtedly that of a polyene and not that of an aromatic molecule.[217]

The dipole moment of fulvene is 1.1 *D* units [218] and the calculated dipole moment[218] agrees well with this result providing one uses a computational scheme more sophisticated than that of the Hückel method which gives much too large a dipole moment both in this case and in general. In this connection, the older valence bond point of view is helpful since an ionic (cross conjugated) structure such as **51** might be expected to have a π-electron dipole moment. But a structure such as **51** would be expected to have a dipole moment of some 10 *D* units, reduced no doubt by back polarisation of the σ-bonds, but to an unknown extent. So the dipole moment evidence for the inclusion of **51** in the valence bond wave function is inconclusive.

The summary on fulvene is that there is virtually no evidence for any aromaticity of this molecule. Only the resonance energy and perhaps the n.m.r. spectrum show slight indications of aromaticity.

4.8 Pentalene (4)

The non-benzenoid hydrocarbon pentalene (**4**) was first reported[221] in 1973. The earlier failures to prepare this molecule stand in interesting contrast to the plethora of theoretical predictions about its properties.

The electronic absorption spectra of the 1-methyl[221a] and 1,3-dimethyl derivatives[221] are known. The molecule is most unstable, being prepared trapped in a matrix at -196°C, and this was inferred earlier from the fact that the diketone (**52**) shows no sign of enolising.[222] The bis (dimethylamino) derivative (**53**) is known and may owe its stability to the formation of a 10-π electron cycle with each amino group adding one electron to the 8-π electrons already present in the ring, particularly since the pentalene dianion (**31**) is known.[224]

4a 4b

Figure 46

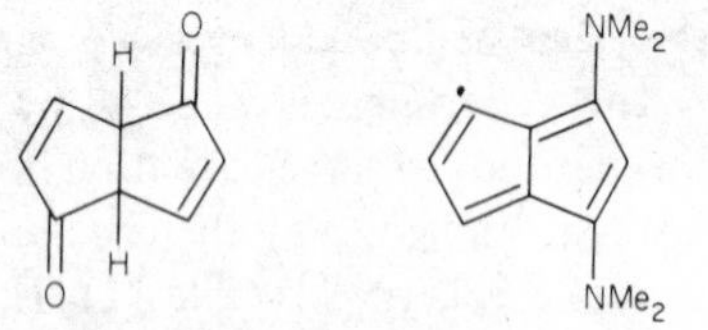

H O H O 52 NMe₂ NMe₂ 53

Figure 47.

On the theoretical side, the older theories suggested that pentalene might be quite a stable molecule. It has two Kekulé structures like benzene and its Hückel resonance energy is sizeable (*cf* table 11). There is nothing in this information to suggest extreme instability. Nor should the ring strain in the two five-membered rings be too great. On the other hand, a small change in the definition of the reference molecule leads to the result that the molecule is non-aromatic or even antiaromatic.[224a]

Craig's rule[46] shows that the ground state, if built as a linear combination of (4a) and (4b), will not have symmetry A_g but rather B_{1g} of D_{2h}. The implication of this result is uncertain. Another suggestion[225] is that pentalene is unstable towards a vibrational motion which creates two acetylene and one diacetylene molecule as in figure 48. It has also been suggested,[226] using a Hückel type theory

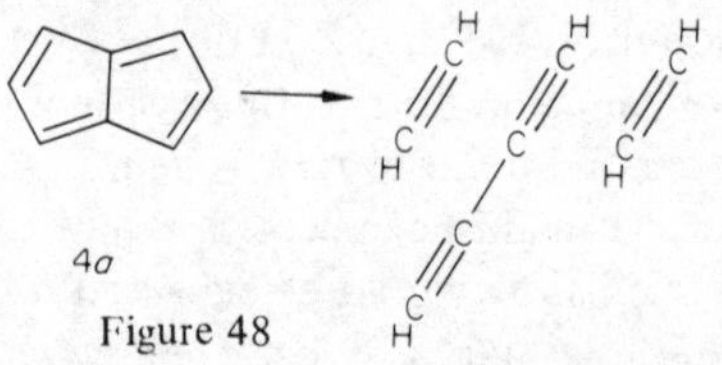

4a

Figure 48

with self-consistency procedures for both the resonance integral and the coulomb integral, that pentalene has five and not four bonding molecular orbitals and so will pick up two electrons to form the known[224] pentalene dianion $(C_8H_6)^{2-}$. The experimental evidence does not lead rigorously to this conclusion, but it does seem an intuitively reasonable inference.

More refined π-electron calculations[176] using a parabolic potential function to represent the σ-electrons (*cf* chapter 3.8) suggest that pentalene is more stable, by about 1 eV, in C_{2h} symmetry than in D_{2h} symmetry. Under the same theoretical procedure, azulene correctly retains its C_{2v} symmetry, although a C_s structure is also of low energy. If the molecule does show marked bond length alternation in the C_{2h} structure its magnetic susceptibility will not be the high paramagnetic value predicted[227] for the D_{2h} structure.

It should be clear by now that no certainty attaches to the calculations which have so far been made on the pentalene molecule and more experimental evidence is needed in this area.

4.9 Summary

In this section are seven examples of how to apply both theoretical and experimental information side by side in discussing the aromaticity problem.

Table 15. *Diamagnetic susceptibility,* χ_m, *susceptibility* exaltation, Λ, and diamagnetic anistropy ΔK *of some compounds of chapters 4 & 5*

Molecule	χ_m (obs)	χ_m (calc)	Λ	ΔK (obs)
azulene	91.0[a]	64.5	26.5	–
p-benzoquinone	40.0[b]	40.9	–0.9	42.4[c]
2, 6-dimethyl pyran -4-one	59.5[d]	62.65	–3.1	–
ferrocene	125[e]	*	–	56.0[f]
fulvene	42.9[g]	41.9	1.0	–
phenanthrene	129[h]	81.7**	47.3	165[i]
phthalocyanin	–	–	–	770[h]
4-pyrone	–	–	–	22.9[j]
thiophen	57.5[k]	44.4***	13.1	37.2[l] 50.1[m]
tropolone	61.0[a]	52.0	9.0	–
tropone	54.3[n]	51.0	3.3	–

All values in units of -10^{-6} cm^3 mol^{-1} χ_m (calc) is calculated from Haberditzl, W. (1966) *Angew, Chem. (Internat. Ed.)* **5**, 294 (table 11)

* The contribution of the iron atom is unknown.

** Using an annelation increment of 20 (Dauben, Jr., H.J. Wilson, J.D. and Laity, J.L. (1969) *J. Amer. Chem. Soc.*, **91**, 1991

*** Ref. 7. **vol.2.** p. 184

References to table 15

a. Klemm, W. (1957) *Ber.* **90,** 1051
b. Krishnan, K.S. and Banerjee, S. (1935) *Phil. Trans,* **234A,** 265
c. Le Fèvre, R.J.W. and Murthy, D.S.N. (1970) *Aust. J. Chem.* **23**, 193
d. Ref. 261
e. Mathis, R., Sweeney, M. and Fox, M.E. (1964) *J. Chem. Phys.* **41** 3652; Wilkinson, G., Rosenblum, M., Whiting M.C. and Woodward R.B. (1952) *J. Amer. Chem. Soc.,* **74,**2125
f. First ref. of reference e above.
g. Thiec, J. and Wiemann, J. (1956) *Bull. Soc. chim. France, 177*
h. Lonsdale, K. (1937) *Proc. Roy. Soc.,* **159A,** 149
i. Ref. 170
j. Benson, R.C., Norris, C.L., Flygare, W.H. and Beak, P. (1971). *J. Amer. Chem., Soc.,* **93**, 5591
k. Foex, G. (1957) Constantes selectionées diamagnétisme et paramagnétisme in *Tables de Constantes et Données Numeriques,* Masson et Cie, Paris.
l. Le Fèvre R.J.W. and Murthy, D.S.N. (1966) *J.Amer. Chem. Soc.* **19**, 1321
m. Sutter, D.H. and Flygare, W.H. (1969) *J. Amer. Chem. Soc.* **91.** 4063
n. Nozoe, T., Mukai, T., Takase, K. and Nagase, T. (1952) *Proc. Japan Acad.* **28,** 447. (1954) *Chem. Abstr.* **48,** 2678

5. Aromaticity of Some Selected Classes of Compounds

Introduction – seven-membered carbocycles – quinoid structures – 4-pyrones and similar molecules – porphin, porphyrins and phthalocyanins – summary

5.1 Introduction

In chapter 4, we discussed some individual molecules whose aromaticity or lack of aromaticity is not in serious doubt. In the present chapter, we discuss four groups of molecules whose aromatic character is in some doubt or which are rarely mentioned in reviews on aromaticity.

One of our purposes here is to show that two molecules may look very much alike in a qualitative sense but may have quite different aromaticities (C_7 cycles), so that membership of a certain general class of compound does not necessarily mean that a given compound is aromatic.

5.2 Seven-membered Carbocycles

In this section, the more recent physical data on the tropylium cation (**6**), tropone (**22**) and tropolone (**54**) (figure 49) are dealt with in order to illustrate a case of a mixed degree of aromaticity within a group of apparently similar compounds. This general area has been reviewed[228] and indexed.[229]

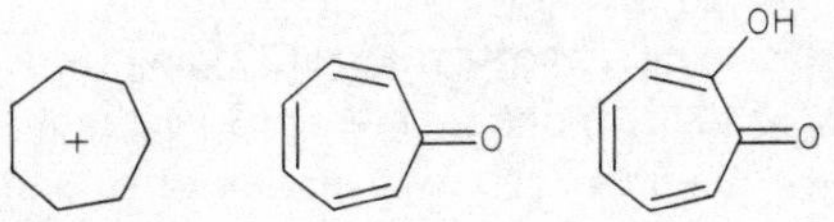

6 22 54

Figure 49.

Some X-ray information on the geometry of various seven-membered carbocycles is shown in figure 50. In all cases, the C_7 ring is planar or very nearly so.

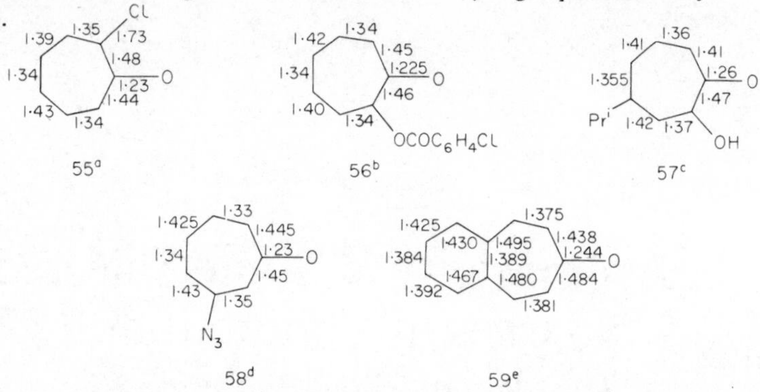

55[a] 56[b] 57[c]

58[d] 59[e]

Figure 50. X-ray diffraction results of some seven-membered carbocycles (bond lengths in Å) (a) Forbes, E.J., Gregory, M.J.. Hamor, T.A. and Watkin, D.J. (1966). *Chem. Comm.* **114** (b) Schaefer, J.P. and Reed, L.L. (1971) *J. Am. chem. Soc.*, **93**, 3902. (c) Derry, J.E. and Hamor, T.A. (1972) *J. Chem. Soc.* Perkin II, 694 (d) Cruickshank, D.W.J. Filippini, G. and Mills, O.S., (1972) *Chem. Comm.* 101. (e) Hata, T., Shimanouchi, H. and Sasada, Y. (1969) *Tetrahedron Letters* **9**, 753.

The 2-chloro (**55**), the 3-azido (**58**) and the 4,5-benzo-derivative (**59**) of tropone all show marked carbon-carbon bond length alternation in the seven-membered ring. Similar results occur for the *iso*propyl (**57**) and *p*-chlorobenzoyl derivative (**56**) of tropolone. If tropone itself has similar bond length alternation then it seems fair to infer that the wave function of tropone is well represented by the single Kekulé structure (**22**) so that there is only a small amount of cyclic delocalisation of the π-electrons.

The electron diffraction experiments on tropone (**22**)[230] and tropolone (**54**)[230a] give results which are, for our present purposes at least, ambiguous. The X-ray work[62] on tropylium salts which was mentioned earlier in this article (chapter 2.2) is as yet far from conclusive evidence on the geometry of the tropylium cation.

There is thermochemical information available for several C_7 carbocycles. Taking the tropylim cation (**6**) first, its gas phase resonance energy is estimated[231] at 58 kcal mol^{-1} and this is considerably larger than that (36 kcal mol^{-1}) of benzene. This result is supported by the fact that the benzyl cation isomerises into the tropylium cation in the mass spectrometer.[232] This rearrangement is usually considered as good evidence for the aromaticity of the tropylium cation. It should be recalled, however, that other factors such as vibrational energy may well be important in differences of this kind between two ions. (*cf* chapter 2.2).

The energetics (tables 4 and 12) of tropone (**22**) and its derivatives point to a resonance energy for the tropone ring of ~ 13 kcal mol^{-1}. The experimental evidence is the heat of hydrogenation of tropone itself which, together with assumptions as to the model compounds, gives a resonance energy of 11.9 kcal mol^{-1}, while the combustion experiment yields ~ 13.8. The heat of combustion of 2,7-dimethyl-4,5-benzotropone (**60**) leads to a resonance energy for this molecule of 49 kcal mol^{-1} and if we allow 36 kcal mol^{-1} for the benzene ring, we have some 13 kcal mol^{-1} for the seven-membered ring.

This figure of about 13 kcal mol^{-1} for the resonance energy of tropone compares with that of 9.0 kcal mol^{-1} for tropylidine (cycloheptatriene) (**61**) and 13.2 kcal mol^{-1} for heptafulvene (**5**). It seems then that tropone is not aromatic. But tropone is flat while

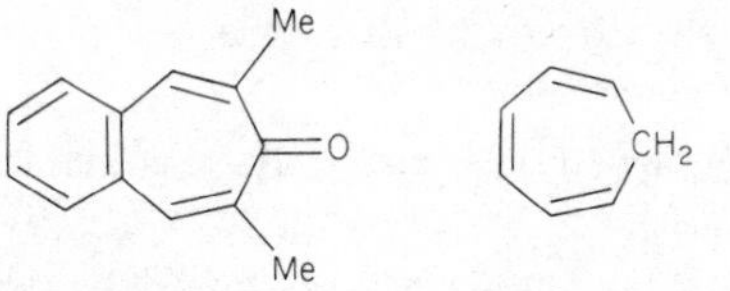

Figure 51

tropylidine (**61**) is tub-shaped[232a] and there is no apparent reason, non-bonded interactions apart, why tropone should not also be tub-shaped unless the π-electrons hold the molecule planar. So perhaps this planarity indicates some slight aromatic character.

The heat of combustion (table 12) for tropolone (**54**) and 2-aminotropone leads to a classical resonance energy of 19 and 18 kcal mol^{-1} respectively. The strong hydrogen bond in tropolone and 2-aminotropone may well contribute some 5 kcal mol^{-1} so the resonance energy of the ring is ~ 14 or 13 kcal mol^{-1} and similar to that obtained above for tropone.

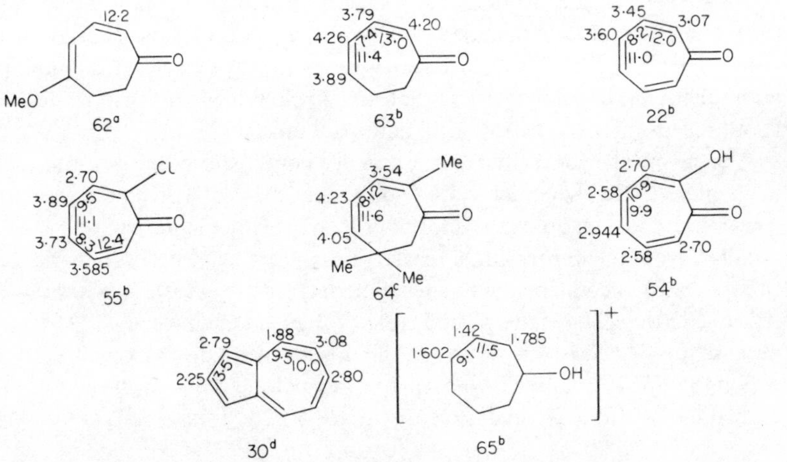

Figure 52. N.m.r. data for some seven-membered carbocycles. The chemical shift of a proton is given by the position of the proton and ^3J values (Hz) are given along the C-C bonds linking the vicinal protons.
(a) Chapman, O.L. (1963) *J. Am. chem. Soc.* **85**, 2014. (b) Ref. 181. (c) Bothner-By, A.A. and Moser, E. (1968) *J. Am. chem. Soc.* **90**, 2347. (d) References 179-182.

The proton chemical shifts and ^3J coupling constants of some seven-membered carbocycles are summarised in figure 52. The data for azulene and for some partly saturated C_7 compounds are included for comparison. Tropone(**22**) and its 2-chloro-derivative (**55**) show a small downfield shift as compared with the partially saturated reference compounds (**63, 64**). Tropolone (**54**) shows larger downfield shifts, comparable with those of azulene (**30**). The downfield shifts of the hydroxytropylium cation (**65**) are even larger but there are of course complications in dealing with ions and it is also true that the chemical shifts of many of this series of compounds[181] are sensitive to the nature of the solvent in which they are measured. Too much emphasis should not be placed on these results dealing with the chemical shifts alone.

The ^3J coupling constants, however, give much the same result with marked alternation of the constants in tropone and its simple derivatives and much less alternation of the constants in tropolone and the tropylium salts. In addition, these ^3J coupling constants are less sensitive[181] to solvent than are the chemical shifts.

The diamagnetic exaltation, Λ, is 3.3 (table 15) for tropone. The tropolone Λ value is rather larger (9.0) and this suggests a small degree of aromaticity, if we recall that the Λ value of benzene is 14 and that of azulene is 26.5.

Thus overall, the magnetic evidence is in line with the earlier geometrical and thermodynamic evidence, all agreeing that tropone is virtually non-aromatic, tropolone is, only slightly aromatic and the tropylium cation is strongly aromatic.

As to the other spectroscopic evidence, the low position (1590 cm^{-1}) of the carbonyl stretching frequency in the infrared region[233] has sometimes been taken as evidence for the importance of the tropylium oxide structure (**23**) in the ground state of the tropone molecule, but it is now understood that the 'true' position of this frequency is masked by complicated coupling[234] with other vibrational modes and that no reliable information can be deduced about the aromaticity of tropone from this source.

The chemical evidence on these C_7 molecules illustrates the point made in chapter 2.3 that when there are two species involved in an equilibrium or a chemical reaction, no direct information about either alone can be obtained from the experimental results. Thus, tropone is markedly basic[80] ($pK_a = -0.6$) as compared with eucarvone(**64**) ($pK_a = -4.9$). This basicity tells one nothing about the ground state of the tropone molecule (**22**) because it is evidently due[235] to the stability of the hydroxytropylium cation (**65**). Similarly, the high basicity of tropolone is due to the aromaticity of the 1,2-dihydroxytropylium

+1·20 +0·94 OH
+1·20
OH

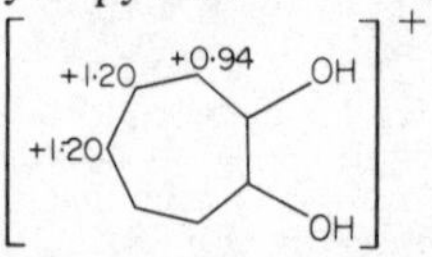

57

Figure 53. Chemical shifts for the ring protons of the 1,2-dihydroxytropylium cation (Olah, G.A. and Carlin, M. (1968) *J.Am.chem. Soc.* **90**, 938).

cation (**57**), as is demonstrated by the 'aromatic' bond lengths[235a] of the C_7 cycle and by the downfield chemical shifts shown in figure 53.

It is more difficult to understand the acidity[236] of tropolone (pK_a = 6.7) as compared with that of acetic acid (pK_a = 4.8) and that of phenol (pK_a = 10.0). This presumably derives from the stability of the anion but no clear explanation of this result is known to us.

Further significant information can be deduced from the dipole moments of these C_7 cycles. Some experimental values are reproduced in figure 54. It is sometimes argued that the large dipole moment (4.30 *D*) of tropone is due to the presence in the ground state wave function of tropone of the

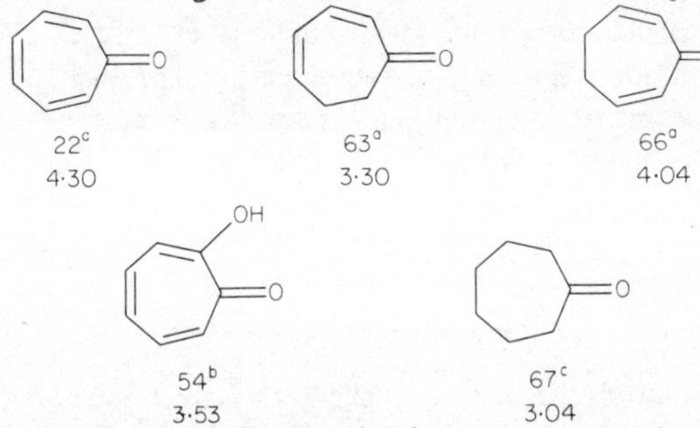

Figure 54. Dipole moments (n Debye units) for some seven-membered carbocycles (a) Ref. 235. (b) Kurita, Y. (1954). *Sci. Reports Tohoku Univ.* First Ser. **38,** 85. (1955) *Chem. Abstr.* **49,** 9989. (c) Gunthard, H.H. and Gaumann, T. (1951) *Helv. Chim. Acta* **34,** 39.

tropylium oxide structure (**23**). In fact, a moment of this size could come from many different sources in a molecule as large as this.

A novel case of a potentially aromatic C_7 carbocycle is the carbene cycloheptatrienylidene[237] (**68**). If the two non-bonded electrons are in a σ-orbital, then there is an empty $2p\pi$ atomic orbital which is free to contribute to the six-π electron seven-π atomic orbital arrangement as in the tropylium cation. This molecule is of course a singlet. But if the two electrons are in separate orbitals, one σ and one π, then the molecule will presumably be a triplet with the π-electron delocalised around the ring in a high energy or antibonding molecular orbital. The chemical and physical properties of such an entity should be of interest.

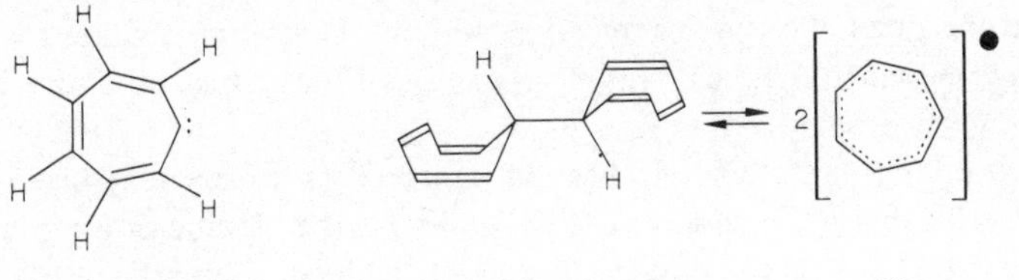

Figure 55.

Another interesting report concerns the tropenyl radical (**70**) and suggests that this molecule has considerable stability. The e.s.r. spectrum[238] shows eight equally spaced lines whose intensities are consistent with the existence of seven equivalent protons. There is however the difficulty that rapid equilibration among seven different structures of classical (non-planar?) nature could give the same experimental result. On the other hand, there is further evidence[231] that the resonance energy of the tropenyl radical (**70**) is some 31 kcal mol^{-1}, much the same as that of benzene. The measurement which gives this remarkable result is a simple decomposition of bitropenyl (**69**) as shown in figure 55. Unless there is some hidden complication or error in this result, it is difficult to see how this value of the resonance energy can be so high and once again we are warned that our understanding of aromaticity is in reality quite superficial.

The general conclusion concerning the aromaticity of these C_7 cycles is that the tropylium cation is aromatic, tropolone is perhaps slightly aromatic and tropone is non-aromatic. This summary is a partial reversal of the position of some twenty years ago when all three molecules were felt to be aromatic.[239, 240]

5.3 Quinoid Structures

The quinones and quinoids generally are usually thought to be non-aromatic[241] yet they apparently contain the magic 6π-electrons in a ring, an electron arrangement which is often regarded as characteristic of aromaticity. The geometry[242] of (planar) *p*-benzoquinone (**71**) is shown in figure 56 and this geometry is undoubtedly that of a non-aromatic system as may be seen from the comparison[243] with acraldehyde (72) in figure 56.

1·322
1·477
1·222 O

C 1·36 C
1·45
C 1·22 O

71 Figure 56. 72

The energetic data on *p*-benzoquinone are given in table 12 and lead to a resonance energy of about 4 kcal mol^{-1}, using the classical procedures for calculating resonance energies. This clearly suggests that *p*-benzoquinone is, at most, slightly aromatic.

The proton magnetic shift of *p*-benzoquinone[244] (τ = 3.33) may be compared with that of benzene (τ = 2.73) and with the olefinic protons of cyclohexene (τ = 4.42). The downfield shift of the *p*-benzoquinone protons from olefine protons of cyclohexene will be due in part to the anisotropy of the carbonyl groups rather than to the presence of a ring current.

The $^3J_{HCCH}$ coupling constant[245] (10.0 Hz) is close to that (10.1) of cyclohexene (figure 24) and different from that (7.56 Hz) of benzene (figure 24). The $^4J_{H_3CCH}$ coupling constants of 2-methyl-*p*-benzoquinone have been recorded[245] but, without the determination of their sign and the analogous constants in related molecules[246], it is difficult to deduce anything about the aromaticity of the molecules concerned.

The diamagnetic anisotropy, ΔK, of *p*-benzoquinone is 42.4 (table 15) and comparison of this value with that (39.4) calculated from a group additivity scheme[247] suggests that *p*-benzoquinone is non-aromatic. The small exaltation, Λ of *p*-benzoquinone (table 15) supports this suggestion.

Both the infrared spectrum[248], which is complicated by the occurrence of Fermi resonance, and also the ultraviolet spectrum[249] give no direct evidence on the question of the aromaticity of the molecule.

Chemically, the quinones often behave as α,β-unsaturated ketones. As an example we offer the reaction[250] shown in figure 57 but it must be remembered here that, in addition to the usual hazards of using chemical reactivity as a guide to aromaticity, there is in this case the complication

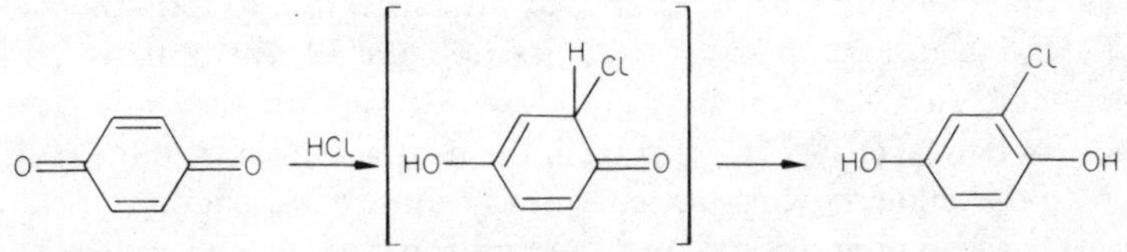

Figure 57.

of the overall aromatisation which occurs in the reaction.

Turning now to other quinoid structures, *p*-quinodimethane (*that is, p*-xylylene) (**74**) is the hydrocarbon analogue of *p*-benzoquinone. If we

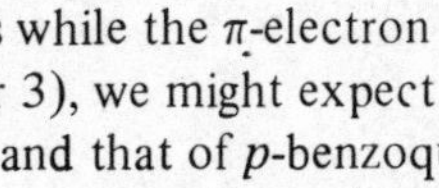

Figure 58.

follow the argument that the σ-electron framework is sensitive to the electronegativity of the atoms while the π-electron framework remains roughly non-polar (*cf* chapter 3), we might expect that the π-electron arrangement of this molecule and that of *p*-benzoquinone would be much the same. *p*-Quinodimethane is only stable in the gas phase[251] or in solution[252] at $-78°$. The electronic spectrum[253] shows a maximum at 274 to 277 nm and it is interesting to notice that the lowest energy π-π transition of *p*-benzoquinone is in just this region (282 nm), supporting the idea that the π-electron structures are similar in the two molecules. The n.m.r. spectrum[253a] supports the quinoid structure.

The planar tetracyano-derivative (**75**) of *p*-quinodimethane is stable in the solid state with the bond lengths[254] shown in figure 59. Again,

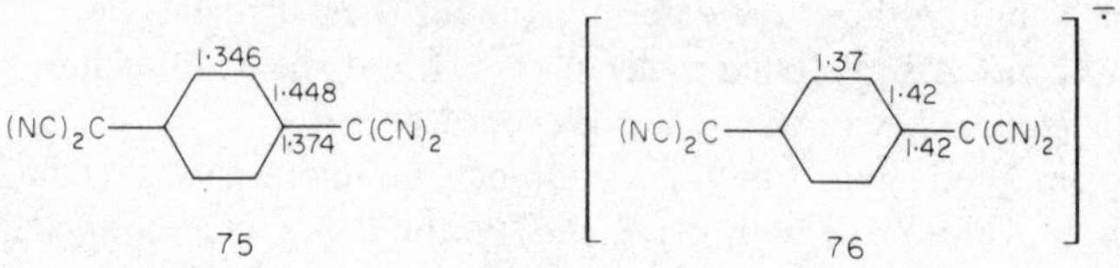

Figure 59.

these bond lengths seem to be typical of a non-aromatic compound.

The heat of combustion (table 12) of the tetracyano compound leads to a resonance energy of 11 or 21 kcal mol^{-1} while tetracyanoethylene has an apparent *negative* resonance energy (table 12) of –33 or –26 kcal mol^{-1}. Evidently the bond energy scheme which underlies the resonance energy idea is breaking down here, since we are uncertain whether resonance energies can be negative (*cf* chapter 7).

Tetracyanoquinodimethane readily accepts an electron to form the radical anion (**76**)[255] whose geometry[256] is shown in figure 59. The central C_6 ring is planar while the CN groups are slightly out of this plane. Evidently there is some geometrical change in going from the neutral molecule to the radical anion with the anion showing the more nearly equal bond lengths.

The over-riding impression from these results is that quinoid molecules are not aromatic.

There are no good computations on *p*-benzoquinone but there are a number of lower quality ones. The internuclear distances are reproduced by the Pariser-Parr-Pople (= CNDO) method applied to the π-electrons alone.[257,258,259] The resonance energy is calculated[259] by a variant of the same method to be about zero.

The conclusion on the quinoid type structures is that, despite the six π-electrons in the ring, the molecules are not aromatic. The only simple explanation of this result that we know of is that the carbonyl groups of *p*-benzoquinone must be treated as individual, tightly bound entities, rather in the 'molecule in molecules' spirit[260] than to break tham up into such structures as those of figure 60. But this does not explain the lack of aromaticity of *p*-quinodimethane (**74**) and the entire problem of why the quinoid molecules lack aromaticity is an

Figure 60.

open question. It is interesting to notice that valence bond theory where there is just one Kekulé structure is more successful than is molecular orbital theory in predicting that quinoid structures are not aromatic.

5.4 4-Pyrones and Similar Molecules

These heterocycles are examples of a molecule with marginal aromaticity. These[261] and their sulphur analogues[262] have been reviewed.

There is rather little direct physical evidence on simple pyrones, such as (**79**) and the only pyrone for which bond lengths seem to be available[263] is the thione (**80**). The thione (**80**) is flat and there is a high degree of bond fixation, *that is*, bond length alternation. There are indications,

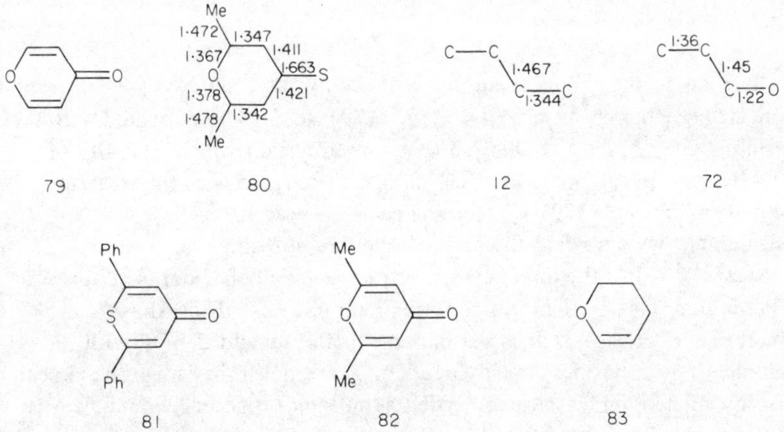

Figure 61.

however, of the shortening of the carbon-carbon single bonds and this is the first sign which would be expected[264] of aromaticity by the bond length criterion.

We have found no data on the heat of combustion of pyrones and only a little indirect evidence as to the resonance energy of the 4-H-thiopyran-4-one (**81**). The resonance energy of the heterocyclic ring in (**81**) has been estimated[265] at 33 kcal mol^{-1} but this old value seems rather large and too much emphasis cannot be placed on a single result. There is n.m.r. evidence[266] that the thio analogues are more aromatic than the simple pyrones. The experimental evidence on the n.m.r. of the 4-pyrones is ambiguous because there is disagreement between the various workers in the field. Some[261] feel that there is substantial evidence for a large ring current in the 4-pyrone (**82**) but their shift data seem to differ from those of others[266,267]. The $^3J_{\mathrm{HCCH}}$ coupling constant[268] in 4-pyrones is about 6.0 Hz while the corresponding vicinal olefinic coupling[269] in 2,3-dihydropyran (**83**) is about 7 Hz, and for simple alkyl substituted 2,3-dihydropyrans is about 6.0 to 6.2 Hz.[270] The value for the fully unsaturated molecule is thus close to that for the partly hydrogenated molecule and this suggests an absence of a ring current in the fully unsaturated molecule. This conclusion is supported by the diamagnetic susceptibility exaltation, Λ, of (**82**) which is close to zero (-3.1, table 15). The diamagnetic anisotropy (**79**) is also small (table 15).

The infrared spectral evidence concerning the carbonyl frequency of the 4-pyrones has been discussed generally[271] but the situation is so complicated that little idea of the importance of the structure (**84**) can be derived from the experimental information.

84 85

Figure 62.

Turning to the chemical equilibria of the pyrones, the exocyclic oxygen is quite strongly basic[272] (pK_a of 4-pyrone (**79**) is -1.14 as compared with say benzaldehyde[273], pK_a -6.99) and this is no doubt due to the stability of the 4-hydroxypyrylium cation (**85**) and has little to do with the structure of the neutral molecule (**79**). Concerning chemical reactivity, there is considerable evidence that the pyran-4-ones are not simple unsaturated ketones[274,275] but the direct evidence **for** aromatic behaviour is in our opinion more meagre than it is generally held to be.[261] Thus, the 2,6-dimethyl derivative (**82**) does not undergo a Michael addition[276] with malonic ester. 4-Pyrones are vinylogues of a lactone so they might be expected to suffer nucleophilic attack somewhat as an ester or lactone would. This is what seems to happen[277] in a complex series of changes which give the overall reaction of figure 63 between 4-pyrone (1 mol) and hydrazine (2 mol). Whether or not the pyrone would undergo typical carbonyl reactions were this type of reaction to be suppressed by magic is, of course, an open question.

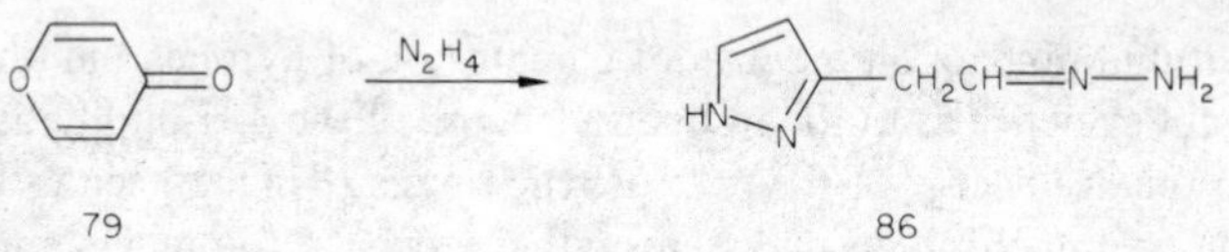

N_2H_4

$CH_2CH{=}N{-}NH_2$

79 86

Figure 63.

The reaction of (**82**) with dimethyl sulphate gives[278] the pyrylium salt (**87**) which is commonly drawn as shown in figure 64. This result

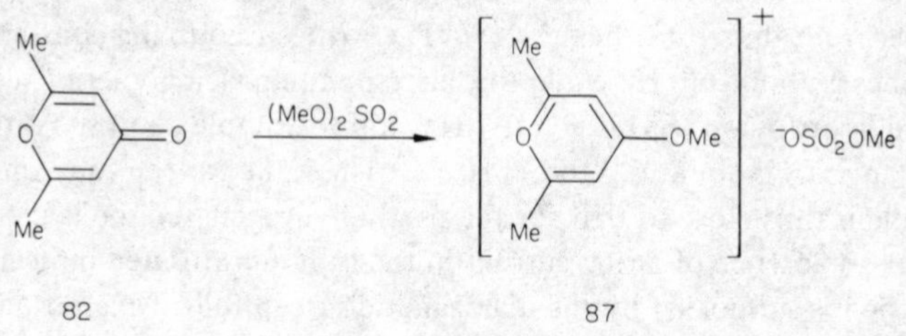

$(MeO)_2SO_2$

$^{-}OSO_2OMe$

82 87

Figure 64.

used to be held to be evidence for the aromaticity of the 4-pyrones but it is now realised that this is merely a reflection of the stability of the aromatic pyrylium cation.

Nitration[275], a typical electrophilic substitution reaction of aromatic molecules, occurs in 4-pyrone **(79)** to give the 3-nitro-derivative. In fact, the yield of the nitro derivative is only 1 per cent and in addition the mechanism is quite unknown so whether or not this result should be considered as evidence for the aromaticity of the 4-pyrone is surely an open question.

An interesting equilibration study[279] is that shown in figure 65 where the equilibrium lies far to the right. The original author feels[279]

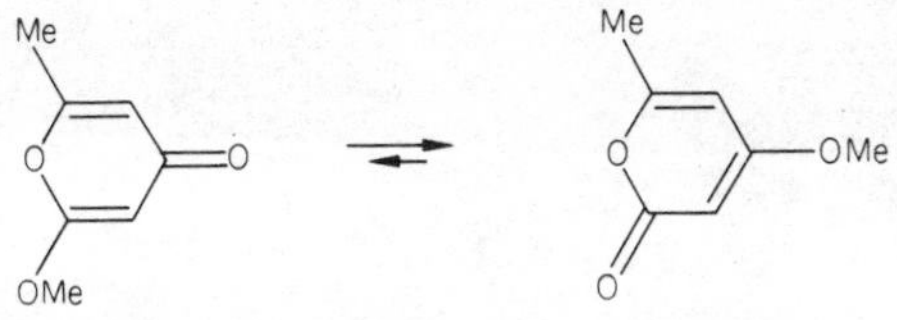

Me Me
O =O ⇌ O —OMe
OMe O
88 89

Figure 65.

that the relative stabilities of **(88)** and **(89)** cannot be said to give a guide to the relative aromaticities of the 2-pyrone and the 4-pyrone and we are in complete agreement with this conclusion.

There are few computations on the 4-pyrones[280] and such as are available are decidedly semi-empirical in nature, using simple Hückel molecular orbital theory with various parameter choices.

In summary, it seems that the evidence for the aromaticity of the pyrones is decidedly weaker than it is generally felt to be. The molecule and its simple derivatives should probably be classed as only slightly aromatic. Interestingly 4-pyrone is isoelectronic with tropone **(22)**, and a similar conclusion was reached for the aromaticity of tropone.

5.5 Porphin, Porphyrins and Phthalocyanins

The parent compound, porphin **(90)**, is shown in figure 66. The four methine bridges between the pyrrolic rings are termed the meso positions.
Porphyrins are porphin **(90)** substituted at *all* the eight β-positions of the four pyrrole rings. In metalloporphyrins the two imino hydrogens are replaced by a single metal atom which is supposed to complex with all four nitrogen atoms. The importance of this group of compounds is that the magnesium complexed porphyrins are involved in photosynthesis (chlorophyll *a* and *b*) and the iron complexes are involved in both oxygen storage and transfer (haems) and also in cell energy transfer (cytochromes). Phthalocyanin **(91)** is the parent of an important class of organic dyes which are formed by a metal atom replacing the two imino hydrogen atoms.

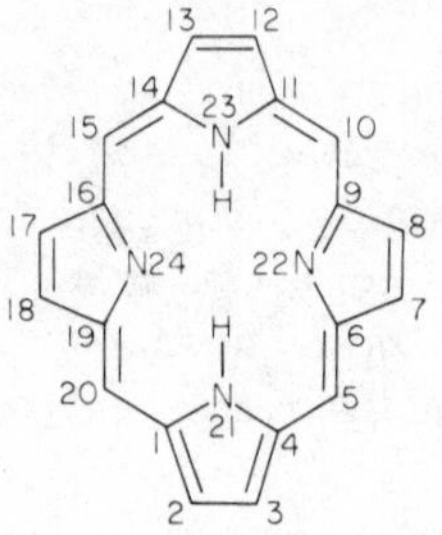

90

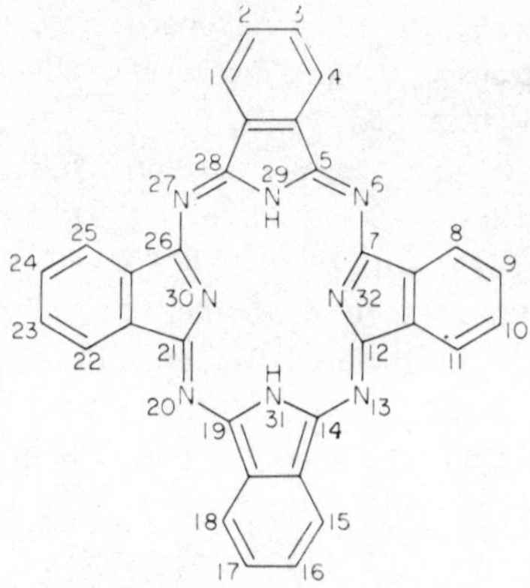

91

Figure 66.

There is a review[281] on some of the physical properties of the porphyrins.

The geometrical information on these molecules shows that the porphin skeleton is 'very nearly planar' in porphin itself[282] while the tetraphenyl derivative[283], with the phenyls in the four meso positions, and its Cu^{II} chelate[284] are both ruffled. The bond lengths of porphin are given in table 13. The table also contains the bond lengths of the molecule

Table 13. *Bond lengths (Å) from* ***X-ray*** *diffraction studies on porphin and haemin*

Bond	Compound				
	Porphin[a]		Haemin[b]		
	max	min	max	min	
Methine	1.398	1.373	1.403	1.352	
$\alpha\beta$ (of pyrrolic ring)	1.465	1.425	1.470	1.434	C_2-C_3 bond of butadiene[c] 1.467
$\beta\beta'$ (of pyrrolic ring)	1.371	1.344	1.346	1.326	
C-N (of pyrrolic ring)	1.389	1.363	1.414	1.359	C-N bond of pyrrole[d] 1.38

a. Ref. 282
b. Ref. 285
c. Ref. 53
d. Bak, B., Christensen, D., Hansen, L. and Rastrup-Andersen, J. (1956) *J. Chem. Phys.* **24**, 720; ref. 55b, M108s.

haemin (also called α-chlorohaemin), which contains substituents in all the eight β positions of the porphin skeleton together with one iron atom and one chlorine atom bound in near the centre of the molecule. The X-ray investigation[285] of this molecule shows that all four nitrogen atoms form a planar square with the iron atom 0.475 Å above the plane of the square and the chlorine atom a further 2.218 Å beyond the iron atom. The pyrrole rings are themselves planar but tilted by 6° or 7° with respect to the plane of the four nitrogen atoms. So the entire molecule forms a saucer shape, convex as viewed from the iron atom.

From the values given in the table, it is clear that the $\beta\beta'$-bonds of both porphin and haemin are close to simple olefine bonds. The $\alpha\beta$-bonds of the two molecules on the other hand, are fairly close to the single bond value as in the C(2)–C(3) bond of butadiene (*cf* table 13). The bonds of the methine bridges of both molecules are close to the aromatic value of 1.40 Å. It seems natural to conclude[266] from this information that there is conjugation through the 16 atom, 18π-electron ring of **(90B)** although it is more common to conjecture that the aromatic

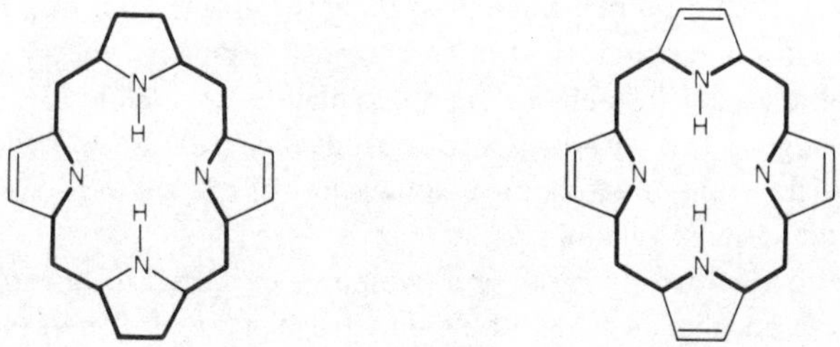

90A

90B

Figure 67.

ring is that of **(90A)** which also has 18 π-electrons but with 18 atoms. The misconception may have arisen from the belief that the $(4n+2)$ rule is *necessarily* obeyed with regard to the number of *atoms* rather than electrons. There is a little evidence for the involvement of the $\beta\beta'$- bonds in the 'great' ring in that the $\alpha\beta$-bond length is perhaps a little short of the 1.47 Å of the central bond of butadiene but the evidence is far from conclusive. If one is convinced by the evidence, then it is reasonable to suppose that there is a few percent of the structure **(90A)** mixed in with **(90B)** in the real structure of the molecule.[287] ^{13}C.m.r. spectra[287a] support structure **(90B)** for the 'great' ring.

In the case of the phthalocyanins, X-ray information is available for the two molecules Cu^{II} phthalocyanin (β)[288] and for Pt phthalocyanin (α)[289] The data are given in the table 14. The $\beta\beta'$-bonds are now aromatic since they are part of the benzene ring but the $\alpha\beta$-bonds are again close to single bonds in length, and the C-N isoindolic ring bonds are like those of pyrrole, and the C-N bridge bonds are shorter and more aromatic looking.

In summary, it seems that the skeletal structure **(90B)** is the preferred single structure for both porphyrins and phthalocyanins. It should of course be realised that the location and the nature of the bonding of the two hydrogen atoms in the centre of the phthalocyanin ring is still uncertain.[290]

Table 14. *Bond lengths (Å) from X-ray diffraction studies on βCu phthalocyanin and αPt phthalocyanin*

Bond	Molecule			
	βCu phthalocyanin[a]		αPt phthalocyanin[b]	
	max	min	max	min
C-N (bridging)	1.335	1.326	1.42	1.26
C-N (isoindolic ring)	1.384	1.351	1.44	1.33
αβ (isoindolic ring)	1.468	1.435	1.53	1.41
ββ′ (isoindolic ring)	1.410	1.390	1.37	1.36

a. Ref. 288.
b. Ref. 289

This means that the nature of the σ-electron distribution which underlies the π-electron distribution is uncertain and we now realise how important a part this σ-electron distribution plays in determining the aromaticity of a molecule.

As mentioned in chapter 1, the thermochemical resonance energy of porphin is 504 kcal mol^{-1} (table 12), and it is plainly impossible to account for a value as large as this without accepting that the structure as a whole is greatly stabilised by unknown factors. Non-empirical calculations on this molecule would be most valuable.

The magnetic information is rather as would be expected for an aromatic molecule. In the porphyrins[291], with no β-hydrogens present, there is a proton signal at low field ($\tau = -1$ to $+1$) due to the four meso protons and another proton signal at high field ($\tau = 13$ to 16) due to the two interior imino protons. This is evidence for a diamagnetic ring current. The large shifts are a reflection that a ring current is proportional[291a] to the area of the ring. Confirmation of this large ring current comes from the diamagnetic anisotropy, ΔK, of phthalocyanin which is 770 (table 15). Recalling that the ΔK of benzene is 58, allowing 4 times 58 for the four benzene rings leaves a value of about 540 for the ΔK of the 'great ring' of the molecule. The modified 'porphin **(92)**[291b] is a good example of the usefulness of n.m.r. information in deciding aromaticity questions.

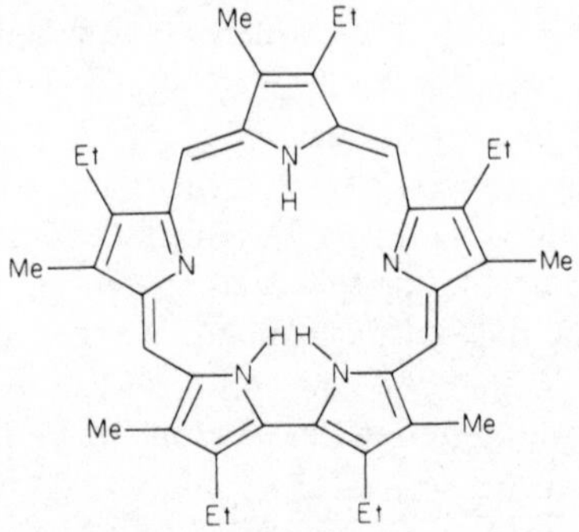

92

Figure 68.

The electronic spectrum of porphin[292] and related molecules[293] is well known and has been discussed theoretically at some length[294].

The chemical reactivity of the molecules of this section fits with the idea that they are aromatic. Thus, bromination of **(90)** with molecular bromine given 2-bromoporphin in 46 per cent yield[295], substitution occurring at a β-position and not at a meso position. If the β-positions are all blocked, say by ethyl groups, chlorination with chlor-sulphuric acid ($ClSO_3H$) at room temperature gives the monomeso-chloro-compound[295]. Formylation of the octaethul Cu^{II} chelate with N, N-dimethylformamide gives[296] the monomeso-formyl derivative in 76 per cent yield. These reactions are those of aromatic rather than aliphatic systems.

The conclusion seems to be that porphin, and the porphyrins and phthalocyanins are aromatic by all the applied standards. It is interesting to notice that however one draws the 'great ring', it contains 18, 22 or 26 π-electrons, $(4n+2)$ in all cases.

5.6 Summary

We have shown that the quinones are not aromatic despite the possession of the 'magic' six π-electrons in a cycle. We have also shown that the porphyrins and the phthalocyanins are aromatic although they differ drastically in their general appearance from benzene itself. In the C_7 cycles, only the tropylium cation is aromatic. The conclusion on the pyrones and related molecules is that these are less aromatic than is commonly supposed.

6. Homoaromaticity

This evocative title was coined by Winstein and his colleagues[297] when they suggested that aromaticity may arise in certain cyclic cations which have neither the σ-electron framework of an orthodox aromatic system nor a parallel $p\pi$ atomic orbital arrangement but which do have some kind of cyclic arrangement of $(4n+2)$ electrons.

There are two related but different ideas here, that of homoconjugation and that of homoaromaticity. The basic idea of homoconjugation can be shown schematically by taking a carbonium ion attached to an allyl group (**93**) and supposing that the π-electron arrangement is like that of the allyl cation (**94**). This is done because there is evidence for interaction between the double bond and the C+ in (**93**). In the homoaromaticity case the π-electron situation

93 94 95 96

Figure 69.

in (**95**) resembles that of the cyclopropenyl cation (**96**). If the latter molecule does show aromaticity, defined by Winstein[297] as increased delocalisation energy of the cyclic as compared with the acyclic analogue, then one might suppose that (**95**) will show some homoaromaticity. A closely related pair of molecules (**97, 98**) perhaps show some signs of homoconjugation[298] [but see discussion in chapter 2. 4. 6].

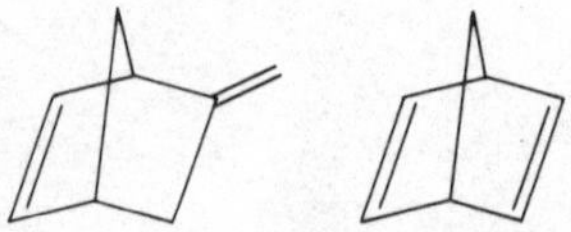

97 98

Figure 70

The above examples (**93, 94, 95**) are not the most general case bcause they use only parallel $p\pi$ atomic orbitals whereas many examples of homoaromaticity involve $p\pi$ atomic orbitals which are not parallel (see below).

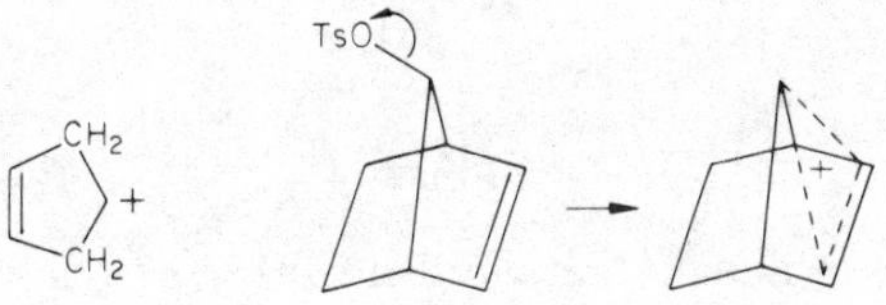

99 101 100

Figure 71.

It is easy to extend this basic idea to more complicated examples. The bishomoaromatic situation is shown schematically in (**99**) with the explicit example of the celebrated norbornenyl cation (**100**) which is generated[299] from (**101**). A similar example[300] is the bishomofulvene type structure (**102**).

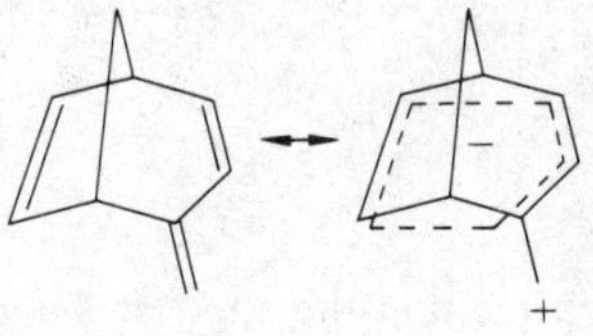

102

Figure 72.

The trihomoaromatic situation is shown in (**103**) which is generated from (**104**).

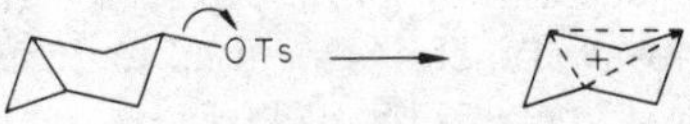

104 103

Figure 73.

A crucial experiment in the development of the idea of homoaromaticity was the solvolysis of the **anti**-7-norbornenyl tosylate (**101**) which is about 10^{11} times as fast as that of the saturated analogue. The corresponding energy difference is about 15 kcal mol^{-1} in free energy of activation and this is a large amount of energy in terms of normal substituent effects. It was to explain this that the homoaromatic ion (**100**) was invoked as an intermediate.[299]

This interpretation is not universally accepted and there are some basic problems with it. One question here is of how much the double bond would affect that cation without delocalisation of the π-electrons of the double bond. It may be the case that the classical formulation (**105**) is a better representation of the true wave function of the norbornenyl cation.

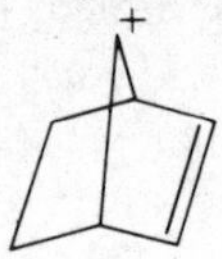

105

Figure 74.

This latter attitude of mind is that of other workers, notably H.C. Brown and his colleagues.[301] They maintain that observations made on homoaromatic cations are really made on rapidly equilibrating mixtures of classical ions. Thus, in the norbornyl cation which is formulated in a homoaromatic sense as in **(106)** they envisage an equilibrium between the classical ions **(107a)** and **(107b)**.

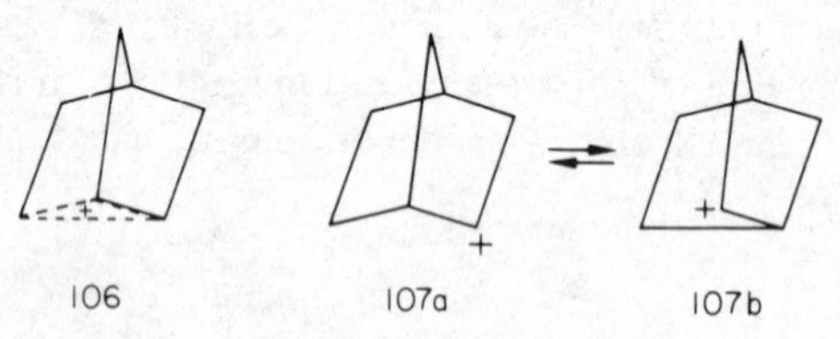

Figure 75.

It seemed likely at one time that the n.m.r. evidence[302] would be helpful in deciding between the two possibilities but in fact the time scale of this experiment is so long (10^{-3} s) that the equilibration may take place faster than this and no conclusion is possible.

There are two difficulties in this area whose existence is not widely appreciated. The first is that representing the electron organisation in molecules by pictures such as **(100)**, **(105)** *etc* is a vague and imprecise statement. Only when wave functions are used to describe the electron organisation can exact statements be made. In particular, different workers may well mean different things by the same picture and this generates polemics. If it is objected that wave functions are too complicated to be generally useful, then pictures of the electron organisation in terms of overlap populations may help.[303] But the nuclear geometry remains a critical (and largely unknown) factor.[304]

The second point which is more widely appreciated is that much of the evidence which is available on this problem of homoaromaticity is derived from chemical reactivity studies and in particular it is sometimes argued that the nature of the products tells one something about the nature of the precursors. For example, the presence of the ether **(108)** among the reaction products of the solvolysis of **(101)** is held to be evidence that the ion **(109)** was present at some stage. The conclusion does not follow from the evidence since quite complex changes may occur between the intermediate and the product.

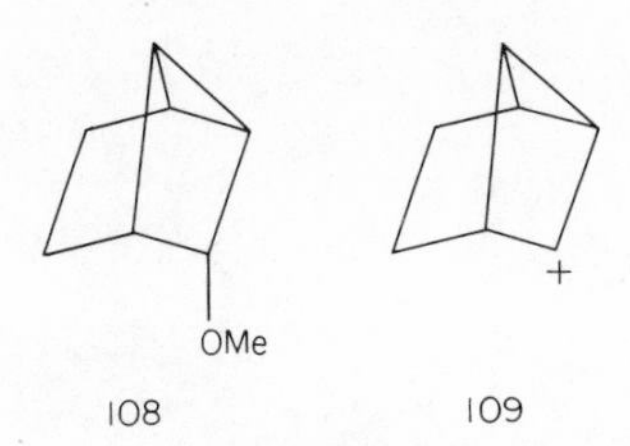

Figure 76.

A particularly interesting result would be the possible demonstration of homoaromaticity in a stable neutral molecule where external complications are minimised and the natural choice for such a test is the molecule tropylidene (**61**) which may be represented as monohomobenzene (**110**). So far, the only result on this molecule seems to be that there is some exaltation of the diamagnetic susceptibility when $\Lambda = 8.1$ (table 6)

Figure 77.

indicating a possible ring current.

The e.s.r. evidence is often held[305] to give information on the homoaromaticity controversy but it is relevant for radicals only, even then the reasoning involves numerous assumptions whose validity seems doubtful.

Attention should also be drawn to discussions of the homotropylium cation[306] (**111**), and the homoaromatic anion[307] (**112**).

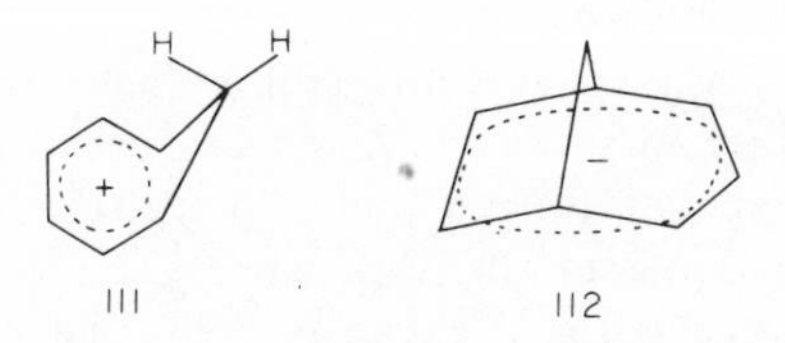

Figure 78.

7. Antiaromaticity

Certain molecules[308,309,310] have come to be described as 'antiaromatic' which means that the molecule is destabilised by the delocalisation of the π-electrons in the same way as a conventional aromatic molecule is thought to be stabilised by the delocalisation of the π-electrons. As an example, we take the cyclopropenyl cation (**7**) which is stabilised by the delocalisation of the π-electrons and compare

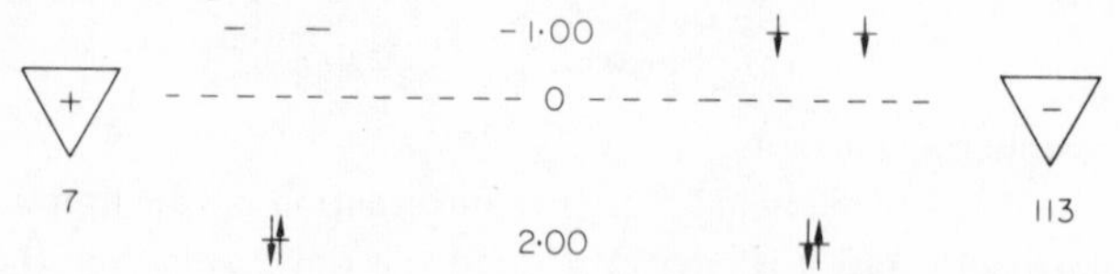

Figure 79. Vertical axis orbitals energy in units of β (ref. 28, p.44).

it with the cyclopropenyl anion (**113**) which is destabilised by the delocalisation of the π-electrons. The π-electron energy levels as calculated[28] from simple Hückel theory are shown in figure 79, assuming that the carbon atoms form an equilateral triangle. We may generalise this result by suggesting that any $4n$ cycle of π-electrons will be antiaromatic. An explicit chemical example[308] is that the cyclopropenyl anion is formed only with difficulty and is 'unstable'. The cyclopentadienyl cation is another such case.[311]

At first sight, the term antiaromatic is an attractive and useful term. But reflection raises the question that, if the delocalisation of the π-electrons is going to raise the total energy, why should the π-electrons delocalise at all? Why should not they simply revert to the structure (**114a**) shown in figure 80? Notice carefully that the geometry of such an ion is not the same as that of (**113**).

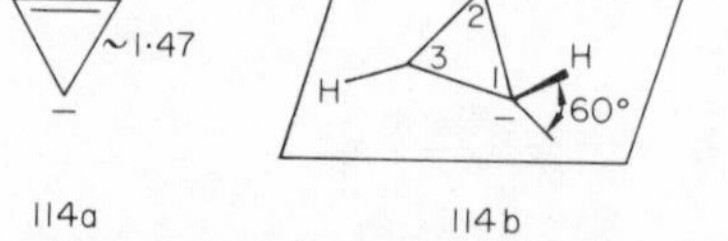

~1·34 ~1·47 114a H 2 3 1 H H 60° 114b

Figure 80. Bond lengths in Å.

This question of antiaromaticity raises several interesting and interconnected problems, some of which have been mentioned in the theory section. One problem is whether or not one should try to use a reference molecule with which to compare the molecule under discussion. It is always important to bear in mind that **all** the electrons, σ and π together, will arrange themselves and the nuclei in such a way as to minimise the total energy of the whole array of electrons and nuclei. A second point is that both σ *and* π-electrons are important in determining the aromaticity as discussed in the theory section. A third point is that our molecular orbital wave functions are still quite crude representations of the true wave function and so one must not put too much reliance on theoretical results.

A more subtle point is that raised in chapter 3.2. This is the common misunderstanding that molecular orbitals are unique. We now know, from the work in chapter 3.2, that the individual molecular orbitals are not unique and it is interesting to see what the significance of this point is in the cyclopropenyl anion. The wave function Ψ of the Kekulé form of the ion (**114a**) is approximately

$$\Psi = \left| \pi_a \frac{(\pi_b + \pi_c)}{(2)^{1/2}} \pi\bar{\pi}_a \frac{\overline{(\pi_b + \pi_c)}}{(2)^{1/2}} \right|$$

where π_a, π_b and π_c stand for the $2p\pi$ atomic orbitals on atoms *a, b* and *c* and the bar above the orbital denotes β-spin, no bar α-spin. Taking the sum and difference as before (chapter 3.2) gives the *same* many-electron wave function in the form.

$$\Psi = \left| \left[\frac{\pi_a}{2^{1/2}} + \frac{\pi_b + \pi_c}{2}\right] \left[\frac{\pi_a}{2^{1/2}} - \frac{\pi_b + \pi_c}{2}\right] \left[\frac{\overline{\pi_a}}{2^{1/2}} + \frac{\overline{\pi_b + \pi_c}}{2}\right] \left[\frac{\overline{\pi_a}}{2^{1/2}} - \frac{\overline{\pi_b + \pi_c}}{2}\right] \right|$$

This form is closely related to the usual delocalised molecular orbital form, although not quite identical with it. The two higher energy electrons are in one molecular orbital and not in two molecular orbitals. The fact that these delocalised molecular orbitals are not quite identical with the Hückel ones is perhaps due to the different nuclear geometries of (**113**) and (**114a**).

This attitude of mind is borne out by some recent detailed computations[312] on the C_3 anion. The lowest energy configuration of this system is an isocelles triangle with the unique hydrogen on C_1 60° out of the plane (**114b**) and the populations show almost one π-electron on each of carbons 2 and 3 and 1.51 π-type electrons on C_1. The situation at C_1 is complicated by the out of plane hydrogen atom (H_1) but it seems that there is an accumulation of π-type charge at C_1 rather as in the classical picture of the anion (**114a**). This result is quite unlike that of the antiaromatic system (**113**) with 4/3 π-electrons on each carbon.

This example of the cyclopropenyl anion is instructive in that it shows how different two *drawings* or pictures such as (**113**) or (**114a**) can appear while the corresponding wave functions are virtually identical or differ only in fine points.

Another example[313, 314] of this point is the celebrated cyclobutadiene molecule where the square planar (D_{4h}) molecule is a triplet while the rectangular planar (D_{2h}) molecule is a singlet.

It seems to us that there is only one basis on which the term 'antiaromatic' can have a valid meaning. This is the situation in which the σ-electrons hold the molecule into an equilateral triangle. And before this can be the case, certain inequalities must hold among the energy quantities. To simplify matters, let us suppose that all energy quantities are positive being measured from zero far down on the energy scale. Then, for antiaromaticity to be possible, we must have

$$(E^{\sigma} + E^{\pi})_{113} < (E^{\sigma} + E^{\pi})_{114a}$$

if we are to suppose that the ion will take up the geometry and electron arrangement of **113**. If, at the same time, we suppose that the π-electrons are in an unfavourable arrangement in **113**, we must mean by this that

$$E^{\pi}_{113} > E^{\pi}_{114a}$$

and so we must have

$$E^{\sigma}_{113} < E^{\sigma}_{114a}$$

In words, the σ-electrons insist on the ion being (**113**), and then the term antiaromatic is a meaningful one.

It might seem, however, that we are now in worse difficulties than ever because if we suppose that the situation is generally true in aromatic and antiaromatic molecules, it destroys most if not all of our π-electron theory as we know it today. For example, in benzene is it the σ-electrons which hold the molecule in a regular hexagon and so are the π-electrons important in determining the geometry of the molecule at all ? And if the σ-electrons determine the geometry, is it not also likely that they determine the energetics as well, and are resonance energy arguments primarily concerned with the σ-electrons and not with the π-electrons? If so, the $(4n+2)$ rule disappears and the π-electron energy level diagrams lose their significance. Perhaps this reflection helps to show how important is the accurate all electron computation of the wave function of a molecule.

It might be argued, however, that a simpler definition of an antiaromatic molecule is one with a paramagnetic ring current[87, 104], as revealed by the n.m.r shifts (*cf* chapter 2.4) and, less convincingly, by magnetic susceptibility measurements.[104] But of course the subject of ring currents is itself a matter of controversy.[37,38,94]

Finally, it is interesting to note that the concept of 'antihomoaromaticity' has been proposed to account for the fact that protonation of 1, 6-methano [*10*] annulene (**45**) is difficult to achieve.

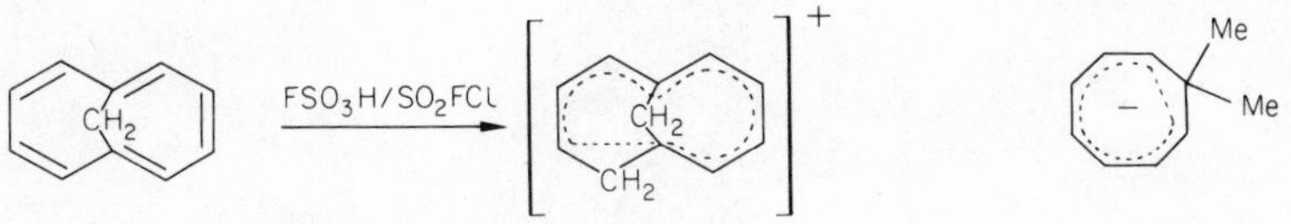

45 115 116

Figure 81

This is attributed[315] to the 'antihomoaromatic' character of the ion (**115**) with its C_9^+ character. Another case[316] is the potentially antihomoaromatic 8 π-electron anion (**116**).

8. Conclusions and Comments

We have discussed many of the definitions of aromaticity which have been proposed over some fifty years. In our opinion, no satisfactory definition yet exists because one can find many exceptions to all the generally known definitions. It is certainly true that no quantitative definition exists now. Some writers have despaired over the concept of aromaticity since it is so vague. It is certainly true that there is a continuous spectrum of compounds from the very to the weakly aromatic and it is usually the latter which cause the difficulty when a decision is necessary as to aromatic or not.

Our choice of definition is the conservative but vague one that 'aromatic' means 'having an electron organisation like that of benzene'. This definition will cover the entire spectrum of aromatic compounds.

One point of the definition is that it is not confined to π-electron systems. It is often supposed that only the π-electrons are involved in determining the aromaticity but this is by no means obviously true.

One important point of the definition is that it assigns a unique position to benzene, and this is not necessarily a good thing. In this context, the purely theoretical definitions are perhaps at their strongest because they do not seem to rely on one unique molecule. But this seeming independence is perhaps misleading, because even the best computations start out with the assumed positions of the nuclei while this is really something which should emerge from the computation. There seems little or no prospect of doing a computation as elaborate as this in the foreseeable future so we must recall that our computations often depend on assumptions about the nuclear positions.

One consequence of the definition of aromaticity is that the original detection of aromatic character by the low chemical reactivity of a reactive looking compound is largely a matter of an accident of history. Presumably this comes about because there is a tendency for transition states to be non-aromatic so that the aromaticity or lack of aromaticity of the ground state is decisive in determining the speed of the reaction (*cf* chapter 2.2.3). If we could find some general proof that transitions states are never significantly aromatic, then we could revive the chemical reactivity criterion of aromaticity.

A number of minor questions concerning aromaticity have been touched on at various points. One intriguing possibility is the existence of aromaticity in an electronic excited state of a molecule which is non-aromatic in its ground state. This was raised in chapter 4.3 in connection with azulene. It would be particularly interesting to compare the aromaticity of the singlet and triplet which arise from the same configuration. At first sight, one is inclined to say that a triplet molecule could not possibly be aromatic but in the present state of our knowledge about aromaticity one should not be dogmatic on this point.

Appendix

Table 16. *Conversion factors.*

Non SI unit	SI equivalent
1 Ångstrom	10^{-10}m (= 0.1 nm)
1 kcal	4.184 kJ
1 Debye (= 10^{-18} esu cm)	1/3 10^{-29} C m
1 cm^{-1} (in infrared spectroscopy)	11.955 J mol^{-1}
1 eV $molecule^{-1}$	96.454 kJ mol^{-1}
1 atomic unit = 1 hartree = 27.21 eV	2624.51 kJ mol^{-1}
1 cm^3 mol^{-1} (in magnetic susceptibility and magnetic anisotropy measurements).	$4\pi 10^{-6}$ m^3 mol^{-1}

Ref. for last entry: Bleaney, B.I. and Bleaney, B. (1957) *Electricity and Magnetism* Oxford University Press, Oxford, p. 652

References

1 Articles in *The Kekulé Symposium*, (1959) Butterworth, London.
2 Bergmann, E.D. and Pullman, B., (eds) (1971) *Aromaticity, Pseudoaromaticity and Antiaromaticity,* Academic Press, New York.
3 *Aromatic and Heteroaromatic Chemistry,* Vol. 1 (1973), Specialist Periodical Reports, Chemical Society, London.
4 Garratt, P.J., (1971), *Aromaticity,* McGraw-Hill, London.
5 Breslow, R., (1965) *Chem. and Engng. News,* **43,** No. 26, 90.
6 Johnson, A.W., (1965) *Science Progress,* **53,** 211.
7 Snyder, J.P., (ed), (1969) *Nonbenzenoid Aromatics,* Vol. 1, Academic Press, London; Vol. 2 (1971).
8 Cresp, T.M. and Sargent, M.V. (1972) in *'Essays in Chemistry',* (eds Bradley J.N., Gillard R.D. and Hudson R.F.), Vol. 4, p. 91. Academic Press.
9 Lloyd, D., (1966) *Carbocyclic Non-benzenoid Aromatic Compounds,* Elsevier, New York.
10 Hafner K., (1964) *Angew. Chem., (Int. Ed.)* **3,** 165.
11 Articles in *Aromaticity,* (1967) Special Publication No. 21, Chemical Society, London.
12 Wheland G.W., (1955) *Resonance in Organic Chemistry,* Wiley, New York.
13 Chandler G.S. and Craig, D.P. (1971) in *'Chemistry of Carbon Compounds',* (ed Rodd) Elsevier, Vol. 3A.
14 See under 'Aromatic Compounds' in recent issues of *"Annual Reports on the Progress of Chem., (B)* The Chemical Society, London.
15 Jones, A.J. (1968) *Rev. pure and appl. Chem.,* **18,** 253.
16 Badger, G.M. (1969) *Aromatic Character and Aromaticity,* Cambridge University Press, London.
17 Garratt, P.J. and Sargent, M.V. (1969) in *Advances in Organic Chemistry, Methods and Results,* **6,** p.1 Interscience, New York.
18 Figeys, H.P., (1969) in *'Topics in Carbocyclic Chemistry',* Vol. 1, 269 Logos Press, London.
19 Harschbarger, W., Lee, G, Porter R.F., and Bauer, S.H. (1969) *Inorganic Chemistry*, **8,** 1683.
20 Paddock, N.L. (1964) *Q Rev. chem. Soc.,* **18,** 168; Cragg, R.H. (1970) in *Essays in Chemistry*, (eds Bradley, J.N., Gilliard R.D. and Hudson R.F.) Vol. 1, p.77. Academic Press, London.
21 Kekulé, F.A. (1865) *Bull. Soc. chim. France,* (2) **3,** 98.
22 Hückel, W. (1955) *Theoretical Principles of Organic Chemistry,* Vol. **1**, 20 Elsevier, London.
23 Pauling, L. (1960) *Nature of the Chemical Bond,* 3rd Edn, Cornell University Press, Ithaca, New York.
24 Armit, J.W. and Robinson, R.(1925) *J. chem. Soc.,* **127,** 1604.
25 Hückel, E. (1931) *Z.Physik,* 70, 204; (1931) **72**, 310; (1938) *Grundzuge der Theorie Ungesattigter und Aromatischer Verbindungen,* Verlag Chemie, Berlin, p.71-85.
26 Reference 23, Chapter 6; Herndon, W.C. (1973) *J.Am. chem. Soc.,* **95,** 2404.
26a Sauter, H., Horster, H-G. and Prinzbach, (1973) *Angew. Chem (Int. Ed),* **12,** 991.
27 Pullman, B. and Pullman, A., (1952) *Les Théories Électroniques de la Chimie Organique,* Masson et Cie, Paris.
28 Streitwieser, A. Jr. (1961) *Molecular Orbital Theory for Organic Chemists,* John Wiley & Sons, New York.
29 Salem, L. (1966) *Molecular Orbital Theory of Conjugated Systems,* Benjamin, New York.
30 Reference 9, p.5.
31 Albert, A. (1959) *Heterocyclic Chemistry,* 201, The Athlone Press, London.
32 *Cf.,* standard organic texts.
33 Peters D. (1969) *J.chem. Phys.* **51,** 1559, 1566.

34 E.g., Cohen, J.B. (1918) *Organic Chemistry for Advanced Students,* Part 2, *Structure,* 2nd Edn. 68 and 411, Arnold, London.
35 (a) Lynden-Bell, R.M. and Harris, R.K. (1969) *Nuclear Magnetic Resonance Spectroscopy,* Nelson, London.
(b) Jackman, L.M. and Sternhell S., (1969), *Applications of Nuclear Magnetic Resonance Spectroscopy in Organic Chemistry* 2nd Edn., Pergamon Press, Oxford, England.
36 Elvidge, J.A. and Jackman, L.M. (1961) *J. chem. Soc.,* **859**
37 Musher, J.I. (1965) *J. chem. Phys.,* **43,** 4081; (1967) **46,** 1219.
38 Abraham, R.J., Sheppard, R.C., Thomas, W.A. and Turner, S. (1965), *Chem. Comm.,* 43.
39 Sondheimer, F. *et al.* (1967), Reference 11, p.75.
40 Reference 12, chapter 8.
41 Peters, D. (1960), *J. chem. Soc.,* **1274.**
42 Reference 16, p. 53 *et seq.*
43 Dewar, M.J.S., Reference 11, p.211.
44 Peters, D. (1963) *J. chem. Soc.,* 2015 and 4017; (1966) (A) 656.
45 Goldstein, M.J. and Hoffman R, (1971) *J. Am. chem. Soc.,* **93,** 6193.
45a Simmons, H.E. and Fukunaga, T.(1967) *J. Am. chem. Soc.,* **89,** 5208.
46 Craig, D.P. ref. 1, p.20: (1959) *Non-benzenoid Aromatic Compounds,* (ed. Ginsburg, D.,) Ch.1 Interscience, New York; Abramovitch, R.A. and McEwen, K.L. (1965) *Canad. J. Chem.* **43,** 2616.
47 Francois, Ph. and Julg, A. (1968) *Theor. Chim. Acta,* **11,** 128.
48a Buenker, R.J., Whitten, J.L. and Petke, J.D. (1968) *J. chem. Phys.* **49,** 2266.
48b Callomon, J.H., Dunn, T.M. and Mills, I.M. (1966) *Phil. Trans. Roy. Soc.* (A), **259,** 499.
48c de Groot, M.S., Hesselmann, I.A.M. and van der Waals, J.H. (1965) *Mol. Phys.,* **10,** 91; Nieman, G.C. and Tinti, D.S. (1967) *J. chem. Phys.,* **46,** 1432; de Groot, M.S. and van der Waals, J.H. (1963) *Mol.Phys.* **6,** 545.
49 Haselbach, E. (1971) *Helv. Chim. Acta.* **54,** 1981.
50 Leroy, G and Jaspers, S. (1967) *J. chem. Phys.,* **64,** 470.
51 Julg, A. and Francois, Ph., (1967) *Theor. Chim. Acta,* **8,** 258.
52 Norman, N and Post, B. (9161) *Acta Cryst.,* **14,** 503: *Cf.,* Rouault, M. and Waziutynska, Z.L. (1957) *Acta Cryst.,* **10,** 804.
53 Haugen, W. and Traetteberg, M. (1966) *Acta Chem. Scand.,* **20,** 1726; Cole, A.R.H., Mohay, G.M. and Osborne, G.A. (1967) *Spec. Acta,* **23A,** 909.
54 Scharpen, L.H. and Laurie V.W. (1965) *J. chem. Phys.* **43,** 2765.
55 *Tables of Interatomic Distances and Configuration in Molecules and Ions,* Special Publications Chem. Soc., London (a) No. 11 (1958) M 196; (b) No. 18 (1965) M 127s.
56 E.g .Holmes, R.R. and Deiters, R.M. (1969) *J. chem. Phys.,* **51,** 4043.
57 Eyring, H., Walter, J. and Kimball, G.E. (1944) *Quantum Chemistry,* p.75, John Wiley & Sons, New York.
58 Reference 556, p.4: Cruickshank, D.W.J. (1960) *Acta Cryst.,* **13,** 774.
59 Sondheimer, F. (1967) *Proc. Roy. Soc.,* **297A,** 173.
60 Oth, JF.M., Woo, E.P. and Sondheimer, F. (1973) *J. Am. chem. Soc.* **95,** 7337.
61 Sundaralingam, M. and Jensen, L.H. (1966) *J. Am. chem. Soc.,* **88,** 198
62 Kitaigorodskii, A.I., Struchkov, Y.T., Khotsyanova, T.L., Vol'pin, M.E. and Kursanov, D.N. (1960), *Izvestiya Akademii Nauk SSSR, Otdelenie Khimicheskikh Nauk,* No. 1, p.39; (1962) *Chem. Abs.,* **56,** 11028; Gould, E.S. (1955) *Acta Cryst.,* **8,** 657.
63 Veillard, A and Del Re, G. (1964) *Theor. Chim. Acta,* **2,** 55; Peters, D., (1963) *Tetrahedron,* **19,** 1539; Randic, M. and Maksic, Z. (1965) *Theor. Chim. Acta,* **3,** 59
64 Refs. 27, 28 and 29: Coulson C.A. (1961) *Valence,* 2nd edn. p. 266, Oxford University Press, London.
65 Reference 28, p.172.
66 Mulliken, R.S., (1955) *J. chem. Phys.*, **23,** 1841; Peters, D. (1964) *Trans. Faraday Soc.,* **60,** 1193.
67 Mulliken, R.S. (1959) *Tetrahedron,* **6,** 77.
68 Landolt-Börnstein Tables, Ed. Schäfer, K. and Lax, E. (1961) 6th Edn., Vol. 2, pt 4, Springer-Verlag, Berlin.
68a Cox, J.D. and Pilcher, G. (1970) *Thermochemistry of Organic and Organometallic Compounds,* Academic Press, London.
68b Stull, D.R., Westrum, Jr. E.F. and Sinke, G.C. (1969) *The Chemical Thermodynamics of Organic Compounds,* John Wiley & Sons, New York.
69 Mortimer C.T. (1962) *Reaction Heats and Bond Strengths,* Pergamon Press, London.

70 Reference 12, Ch. 3.
71 Davis, R.E. and Pettit, R. (1970) *J. Am. Chem. Soc.*, **92,** 716.
72 Peters, D. (1965) *J. chem. Soc.*, 3026.
73 Roberts J.D. (1962) *Molecular Orbital Calculations,* p.48, Benjamin, New York.
74 Turner, R.B., Reference 1, p. 67.
75 Reference 12, p. 89.
76 Reference 69, p. 71.
76a Rodgers, D.L., Westrum, E.F. and Andrews, J.T.S. (1973) *J. Chem. Thermodynamics,* **5,** 733.
77 Fischer, H and Rewicki, D. (1968) *Prog. org. Chem.*, 7, 116, Kosower, E.M. (1968) *An Introduction to Physical Organic Chemistry,* p. 27, John Wiley & Sons, New York.
78 Streitwieser, Jr., A. and Hammons, J.H. (1965) *Prog. Phys. org. Chem.*, **3,** 41.
79 Peters, D. (1959) *J. chem. Soc.*, 1757.
80 Reported by Kende, A. (1965) *Adv. Chem. Phys.*, **8,** 136.
81 (a) Breslow, R. and Yuan, C. (1958) *J. Am. chem. Soc.*, **80,** 5991; (b) Breslow, R., Höver, H. and Chang, H.W. (1962) *J. Am. chem. Soc.*, **84**, 3168.
82 Breslow, R. and Groves, J.T. (1970) *J. Am. chem. Soc.*, **92,** 984.
83 Dixon, W.T. (1970) *J. chem. Soc.*, (B) 612.
84 Reference 16, p. 85.
85 Carrington, A. and McLachlan, A.D. (1967) *Introduction to Magnetic Resonance,* Harper, New York.
86 Waugh, J.S. and Fessenden, R.W. (1957) *J. Am. chem. Soc.*, **79**, 846; Vogel, E., Pretzer, W. and Boell, W.R. (1965) *Tetrahedron Letters,* 3613.
87 Pople, J.A. and Untch, K.G. (1966) *J. Am. chem. Soc.*. **88,** 4811.
88 Reference 35b pt 3. Ch. 6 and pt 4, Ch. 3.
89 Palmer, M.H. (1967) *The Structure and Reactions of Heterocyclic Compounds,* E. Arnold, London.
90 Labarre, J-F., Loth, P. de and Graffeuil, M. (1966) *J. Chim. phys.*, **63**, 460. Labarre, J-F., Crasnier, F. and Faucher, J-P. (1966) *J. Chim. phys.*, **63**, 1088.
91 Nakajima, T and Koda, S. (1966) *Bull. Chem. Soc. Japan,* **39,** 804.
92 Nakajima, T. and Katagiri, S. (1964) *Mol. Phys.*, **7,** 149.
93 Elvidge, J.A. (1965) *Chem Comm.*, 160.
94 Berson, J.A., Evleth, Jr., E.M. and Manatt, S.L. (1965) *J. Am. chem. Soc.*, **87**, 2901.
95 Table 5, Reference g.
96 Kanekar, C.R. and Khetrapal, C.L. (1967) *Curr. Sci.*, **36,** 67: (1967) *Chem. Abs.* **67,** 2666d.
97 Dhingra, M.M., Govil, G., Kanekar, C.R. and Khetrapal, C.L. (1967) *Proc. Ind. Acad. Sci. Sect. A.*, **65,** 203: (1967) *Chem. Abs.* **67,** 63605e.
98 Laslo, P. and Schleyer, P. von R. (1963) *J. Am. chem. Soc.*, **85**, 2017.
99 Cooper, M.A. and Manatt, S.L. (1969) *J. Am. chem. Soc.*, **91**, 6325.
100 Jonathan, N., Gordon, S. and Dailey, B.P. (1962) *J. chem. Phys.*, **36,** 2443.
101 Goldstein, J.H. and Reddy, G.S. (1962) *J. chem. Phys.*, **36**, 2644.
102 Van Vleck, J.H. (1932) *The Theory of Electric and Magnetic Susceptibilities,* Oxford University Press, London.
102a Mitra, S.(1972) in *Transition Metal Chemistry,* ed. Carlin R.L., **7,** 310.
103 Pauling, L. (1936) *J. chem. Phys.*, **4,** 673.
103a Benson, R.C. and Flygare, W.H. (1970) *J. chem. Phys.*, **53,** 4470.
104 Dauben, Jr. H.J., Wilson, J.D. and Laity, J.L. (1968) *J. Am. chem. Soc.*, **90**, 811: (1969) *J. Am. chem. Soc.*, **91**, 1991.
105 Articles in *Radical Ions,* Ed. Kaiser, E.T. and Kevan, L. (1968), Interscience, New York.
106 Reference 16, p. 37.
107 *Cf.* general organic texts.
108 E.g.Thomson, C. (1969) *Annual Reports on the Progress of Chemistry (B)* **66,** p.22. The Chemical Society, London, Allred, A.L. and Bush, L.W. (1968) *J. Am. chem. Soc.*, **90**, 3352.
109 Dewar, M.J.S. (1969) *The Molecular Orbital Theory of Organic Chemistry,* p.274. McGraw-Hill, New York.
110 Wilson, E.B., Decius, J.C. and Cross, P.C. (1955) *Molecular Vibrations: Theory of Infrared and Raman Vibrational Spectra,* McGraw-Hill, New York.
111 Reference 9, p. 104.
112 Reference 16, p. 58.
113 Herzberg, G. (1966) *Molecular Spectra and Molecular Structure,* Vol. 3, ch. 5, Van Nostrand, Princetown, New Jersey.
114 Slater J.C. (1960) *Quantum Theory of Atomic Structure,* Vol. 2 p.7. McGraw-Hill, New York.
115 Watanabe, K. (1957) *J. chem. Phys.*, **26,** 542.
116 Field, F.H. and Franklin, J.L. (1954) *J. chem. Phys.*, **22,** 1895.
117 Baker, C. and Turner, D.W. (1968) *Proc. Roy. Soc.*, **308A,** 19 and References therein: Turner, D.W. (1970) *Molecular Photoelectron Spectroscopy,* Wiley, Interscience, London.
118 Koopmans, T. (1933) *Physica,* **1,** 104.

119 Peters, D. (1966) *J. chem. Phys.*, **45**, 3474.
120 Al-Joboury, M.I. and Turner, D.W. (1964) *J. chem. Soc.*, 4434.
121 Peters, D. (1963) *J.chem. Soc.*, 2019.
122 Reference 9, p. 81.
123 Laurie, V.W. (1956) *J. chem. Phys.*, **24**, 635.
124 Butcher, S.S. (1965) *J. chem. Phys.*, **42**, 1830.
125 Reference 9, p. 26: Clark, D.T. and Lilley, D.M.J. (1970) *Chem. Comm.* 147.
126 Swalen, J.D. and Costain, C.C. (1959) *J. chem. Phys.*, **31**, 1562.
127 *Cf.*, Reference 64, last entry, p. 152.
128 Lucken, E.A.C. (1969) *Nuclear Quadrupole Coupling Constants*, Academic Press, New York.
129 Reference 9, p. 102.
130 Perrin, C.L. (1965) *Progress Phys. org. Chem.*, **3**, 295.
131 Schulman, J.M. and Moskowitz, J.W. (1967) *J. chem. Phys.*, **47**, 3491: Buenker, R.J., Whitten, J.L. and Petke, J.D. (1968) *J.chem. Phys.*, **49**, 2261.
132 Clementi, E. (1968) *Chem. Rev.* **68**, 341.
133 *Cf., Neumann*, D.B. and Moskowitz, J.W. (1969) *J.chem.Phys.*, **50**, 2216 and refs. therein.
134 Peters, D. (1964) *J. chem. Soc.*, 2908.
135 Reference 64, last entry: Murrell, J.N. Kettle, S.F.A. and Tedder, J.M. (1965) *Valence Theory*, Wiley, New York: Flurry, Jr., R.L. (1968) *Molecular Orbital Theories of Bonding in Organic Molecules*, E. Arnold, London.
135a Pople, J.A. (1957) *Quart. Rev. chem. Soc.* **11**, 273: Edmiston, C. and Ruedenberg, K. reference 138a, p. 263: Ref. 33 and refs. therein.
136 Roothaan, C.C.J. (1951) *Rev. Mod. Phys.*, **23**, 69.
137 E.g., Clark, R.G. and Stewart, E.T. (1970) *Q. Rev. chem. Soc.*, **24**, 95.
138 (a) Articles in *Quantum Theory of Atoms, Molecules and the Solid State*, (1966) Ed. Lowdin, P-O., Academic Press, New York: (b) Articles in *Advances in Chemical Physics* (1969) Vol. 14, Interscience, London.
139 Coulson, C.A. (1939) *Proc. Roy. Soc.*, **169***A*, 413: McWeeney, R. (1951), *J. chem. Phys.*, **19**, 1614, (1952) **20**, 920, Mulliken, R.S. (1955) *J. chem. Phys.*, **23**, 1833, 2346.
140 Peters, D. (1963) *J. chem. Soc.*, 2015 and 4017.
140a Christoffersen, R.E. in (1972) *Advances in Quantum Chemistry*, Ed. Lowdin. P-O., Academic Press, Vol. 6, 333.
140b Diercksen, G. and Preuss, H. (1966) *Z. Naturforschung*, **A21**, 863.
140c Praud, L., Millie, P. and Berthier, G. (1968) *Theor. chim. Acta*. **11**, 169.
140d Buenker, R.J. and Peyerimhoff, S.D. (1969) *Chem Phys. Letters*, **3**, 37.
141 *E.g.* (a) Lo, D.H. and Whitehead, M.A. (1969) *J. Am. chem. Soc.*, **91**, 238. and References therein: (b) Chung, A.L.H. and Dewar, M.J.S. (1965) *J. chem. Phys.*, **42**, 756; (c) Dewar, M.J.S. and Gleicher, G.J. (1965) *J. Am. chem. Soc.*, **87**, 685: (d) Dewar, M.J.S. and de Llano, C. (1969) *J. Am. chem. Soc.*, **91**, 789.
142 Boys, S.F. and Foster, J.M. (1960) *Rev. Mod. Phys.*, *32*, 303.
143 Peters, D. (1963) *J. chem. Soc.*, 4017
144 Peters, D., (1963) *J. chem. Soc.*, 2015
145 Dunning, Jr., T.H., Winter, N.W. and McKoy, V. (1968) *J. chem. Phys.*, **49**, 4128.
146 Davies, D.W. (1968) *Trans. Faraday Soc.*, **64**, 2881: Adam, W. and Grimison, A. (1967) *Theor. Chim. Acta*, **7**, 342: Baird, N.C. and Dewar, M.J.S. (1967) *Theor. Chim. Acta*, **9**, 1.
147 Jones, A.J., Alger, T.D., Grant, D.M. and Litchman, W.M. (1970) *J. Am. chem. Soc.*, **92**, 2386.
148 Coulson, C.A. and Altmann, S.L. (1952) *Trans. Faraday Soc.*, **48**, 293.
149 E.g., Eliel, E.L., Allinger, N.L., Angyal, S.J. and Morrison, G.A. (1965) *Conformational Analysis*, Interscience, New York.
150 E.g. Zahradnik, R. Michl, J. and Pancir, J. (1966) *Tetrahedron*, *22*, 1356.
150a Gund, P. (1972) *J. Chem. Ed.*, **49**, 100.
150b Klein, J. and Medlik. A. (1973) *Chem. Comm.* 275.
151 *Cf.*, Parr, R.G. (1963) *The Quantum Theory of Molecular Electronic Structure*, Benjamin, New York, Ch. 3.
152 Pople, J.A., Santry, D.P. and Segal, G.A. (1965) *J. chem. Phys.*, **43**, S129: Pople, J.A. and Segal, G.A. (1965) *J. chem. Phys.*, **43**, S136: (1966) **44**, 3289: Pople, J.A. and Gordon, M. (1967) *J. Am. chem. Soc.*, **89**, 4253.
153 Davidson, R.B., Jorgensen, W.L. and Allen, L.C. (1970) *J. Am. chem. Soc.*, **92**, 749: Sichel, J.M. and Whitehead, M.A. (1968) *Theor. Chim. Acta*, **11**, 220.
154 Bloor, J.E. and Breen, D.L. (1967) *J. Am. chem. Soc.*, **89**, 6835: Kuznesof,

154 (Continued)
P.M. amd Shriver, D.F., (1968) *J. Am. chem. Soc.*, **90,** 1683: Dewar, M.J.S., Harget, A. and Haselbach, E. (1969) *J. Am. chem. Soc.*, **91,** 7521.
154a Hoffmann, R. (1971) *Accounts Chem. Research,* **4,** 1.
154b E.g., Becher, G., Luttke, W. and Schrumpf, G., (1973) *Angew. Chem. (Int. Ed.)*, **12,** 339.
155 Lennard-Jones, J.E. and Turkevich, J. (1937) *Proc. Roy. Soc.*, **158A,** 297.
156 Longuet-Higgins, H.C. and Salem, L. (1959) *Proc. Roy. Soc.*, **251A,** 172: (1960) **257A,** 445.
157 Dewar, M.J.S. and Schmeising, H.H. (1959) *Tetrahedron,* **5,** 166.
157a Peters, D. (1972) *J. Am. chem. Soc.*, **94,** 707.
158 Lo, D.H. and Whitehead, M.A. (1968) *Can. J. Chem.*, **46,** 2027, 2041.
159 Coulson, C.A. and Streitwieser, Jr. A. (1965) *Dictionary of π-Electron Molecular Calculations,* Pergamon Press, London.
160 Wheland, G.W. and Mann, D.E. (1949) *J. chem. Phys.*, **17,** 265: Streitwieser, Jr., A. (1960) *J. Am. chem. Soc.*, **82,** 4123.
161 Smith, W.B., Watson, W.H. and Chiranjeevi, S. (1967) *J. Am. chem. Soc.*, **89,** 1438.
162 Goeppert-Mayer, M. and Sklar, A.L. (1938) *J. chem. Phys.*, **6,** 645.
163 *E.g.*, Reference 4, p. 41.
164 Reference 9, p. 209.
165 Reference 4, p. 138.
166 Trotter, J. (1963) *Acta Cryst.*, **16,** 605.
167 Pritchard, H.O. and Sumner, F.H. (1954) *Proc. Roy. Soc.*, **226A,** 128.
168 Cf., Reference 11 p. 190.
169 Bartle, K.D. and Smith, J.A.S. (1967) *Spec. Acta* **23A,** 1689.
170 Le Fèvre, R.J.W. and Murthy, D.S.N. (1966) *Aust. J. Chem.* **19,** 179.
171 Buu-Hoï, N.P., Mabille, P. and Do-Cao-Thang (1966) *Bull. Soc. chim. France,* 180.
172 Altschuler, L. and Berliner, E. (1966) *J. Am. chem. Soc.*, **88,** 5837.
173 Reference 9, p. 182.
174 Ammon. H.L. and Sundaralingam, M. (1966) *J. Am. chem. Soc.*, **88,** 4794.
175 Wiberg, K.B. (1964) *Physical Organic Chemistry,* p. 120, Wiley, London.
176 François, Ph. and Julg, A. (1968) *Theor. Chim. Acta,* **11,** 128.
177 Reference 9, p. 191.
178 Reference 11, p. 194.
179 Fry, A.J., Bowen, B.W. and Leermakers, P.A. (1967) *J. org. Chem.*, **32,** 1970.
180 Schneider, W.G., Bernstein, H.J. and Pople, J.A. (1958) *J. Am. chem. Soc.*, **80,** 3497.
181 Bertelli, D.J., Andrews, Jr., T.G. and Crews, P.O. (1969) *J. Am. chem. Soc.*, **91,** 5286.
182 Jones, J.A., Alger, T.D., Grant, D.M. and Litchman, W.M. (1970) *J. Am. chem. Soc.*, **92,** 2386.
183 Reference 9, p. 192.
184 Flurry, Jr., R.L. and Bell, J.J. (1967) *J. Am. chem. Soc.*, **89,** 525.
185 Tobler, H.J., Bauder, A. and Günthard, Hs. H. (1965) *J. Mol. Spec.*, **18,** 239.
186 Brown, R.D. and Coller, B.A.W. (1967) *Theor. chim. Acta,* **7,** 261.
187 Anderson, Jr. A.G. and Steckler, B.M. (1959) *J. Am. chem. Soc.*, **81,** 4941.
188 Tyutyulkov, N. and Fratev, F. (1967) *Theor. chim. Acta,* **8,** 62.
189 Reference 16, p. 122.
190 Anderson, Jr., A.G. and Harrison, W.F. (1964) *J. Am. chem. Soc.*, **86,** 708.
191 Evleth, E.M. (1970) *Theor. chim. Acta,* **16,** 22.
192 Borsdorf, R. (1964) *Z. Chem.*, **4,** 422.
193 Masamune, S., Hojo, K., Hojo, K., Bigam, G. and Rabenstein, D.L. (1971), *J. Am. chem. Soc.*, **93,** 4966.
193a Vogel, E., Reference 11, p. 113: *Chimia,* (1968), **22,** 21.
194 Dobler, M. and Dunitz, J.D. (1965) *Helv. chim. Acta,* **48,** 1429.
195 Murrell, J.N. and Hinchliffe, A. (1966) *Trans. Faraday Soc.*, **62,** 2011.
196 Beezer, A.E., Mortimer, C.T., Springall, H.D., Sondheimer, F. and Wolovsky, R. (1965) **J. chem. Soc.,** 216.
197 Davies, D.W. (1959) *Tetrahedron Letters,* No. 8, p. 4.
198 Dewar, M.J.S., Reference 11, p. 195.
199 (a) Calder, I.C., Garratt, P.J., Longuet-Higgins, H.C., Sondheimer, F. and Wolovsky, R. (1967) *J. chem. Soc.*, (C) 1041: (b) Woo, E.P. and Sondheimer, F. (1970) *Tetrahedron,* **26,** 3933: (c) Reference 59, p. 179.
200 Blattmann, H.R., Heilbronner, E. and Wagniere, G., (1968) *J. Am. chem. Soc.*, **90,** 4786.
201 Johnson, S.M., Paul, I.C. and King, G.S.D. (1970) *J. chem. Soc.*, (B) 643.

202 Coates, G.E., Green, M.H.L. and Wade, K. (1968) *Organometallic Compounds,* Vol. 2, Methuen, London.

203 Haarland, A. and Nilsson, J.E. (1968) *Acta chem. Scand.,* **22,** 2653.

204 Cotton, F.A. and Wilkinson, G. (1952) *J. Am. chem., Soc.,* **74,** 5764.

205 Butter, S.A. and Beachell H.C. (1966) *Inorg. Chem.,* **5,** 1820: Fraenkel, G., Carter, R.E., McLachlan, A. and Richards, J.H. (1960) *J. Am. chem. Soc.,* **82,** 5846.

206 Reference 9, p. 89.

207 Reference 9, p. 88.

208 Harris, C.B. (1968) *Inorg. Chem.* **7,** 1517: Armstrong, A.T., Carroll, D.G. and McGlynn, S.P. (1967) *J. chem. Phys.,* **47,** 1104: Hoefflinger, B. and Voitlaender, J. (1963) *Z. Naturforschung,* **18a,** 1065, 1074: (1964) *Chem. Abs.,* **60,** 4811b, c: contributions in *Discussions of the Faraday Society,* **47,** *Bonding in Metallo-organic Compounds,* Faraday Soc., London, 1969.

209 Bak, B., Christensen, D., Hansen-Nygaard, L. and Rastrup-Andersen, J. (1961) *J. Mol. Spec.,* **7,** 58: Reference 55b, M 106s: Harshbarger, W.R. and Bauer, S.H. (1970) *Acta Cryst.* **B26,** 1010.

210 Grant, D.M., Hirst, R.C. and Gutowsky, H.S. (1963) *J. chem. Phys.,* **38,** 470.

211 Abraham, R.J. and Thomas, W.A. (1966) *J. chem. Soc.,* (B) 127.

212 Gerdil, R. and Lucken, E.A.C. (1965) *J. Am. chem. Soc.,* **87,** 213: (1966), **88,** 733.

213 Butler, A.R. and Hendry, J.B. (1970) *J. chem. Soc.,* (B) 170.

213a Linda, P and Marino, G. (1970) *J. chem. Soc.,* (B), 43.

214 Wachters, A.J.H. and Davies, D.W. (1964) *Tetrahedron,* **20,** 2841.

215 Krishnamurthy, V.N. and Soundararajan, S. (1966) *J. org. Chem.,* **31,** 4300.

216 Bielefeld, M.J. and Fitts, D.D. (1966) *J. Am. chem. Soc.,* **88,** 4804: Clark, D.T. and Armstrong D.R. (1970) *Chem. Comm.* 319.

217 Bergmann, E.D. (1968) *Chem. Rev.,* **68,** 41.

218 Julg, A. and François. P. (1967) *Theor. chim. Acta,* **8,** 249 and References therein.

219 Dewar, M.J.S., Reference 11, p. 196.

220 Meuche, D., Neuenschwander, M. Schaltegger, H. and Schlunegger, H.U., (1964) *Helv. Chim. Acta,* **47,** 1211.

221 Hafner, K., Donges, R., Goedecke. E. and Kaiser, R. (1973) *Angew. Chem. (Int. Ed.)* **12,** 337.

221a Bloch, R., Marty, R.A. and de Mayo, P. (1972) *Bull. Soc. Chim. France,* **5,** 2031.

222 Dauben, Jr., H.J., Jiang, S.H-K., and Ben, V.R. (1957) *Hua Hsüeh Hsüeh Pao* **23,** 411: (1958) *Chem. Abs., 52,* 16309h.

223 Hafner, K., Bangert, K.F. and Orfanos, V. (1967) *Angew. Chem. (Int. Ed),* **6,** 451.

224 Katz, T.J. and Rosenberger, M. (1962) *J. Am. chem. Soc.,* **84,** 865.

224a Milun, M.Sobotka, Z. and Trinajstić, N. (1972) *J. org. Chem.,* **37,** 139.

225 Longuet-Higgins, H.C., Reference 1, p. 17.

226 Boyd, G.V. (1966) *Tetrahedron,* **22,** 3409.

227 Nakajima, T. (1971) Pure and Appl. Chem. **28,** 219.

228 Reference 9, chapters 5 and 6.

229 Miller, J. (1968) *Reaction Mechanisms in Organic Chemistry,* Monograph 8, p. 298. *Aromatic Nucleophilic Substitution,* Elsevier.

230 Kimura, K., Suzuki, S., Kimura, and Kubo, M. (1957) *J. chem. Phys.,* **27,** 320: ref. 55b, M 138s.

230a Kimura, M. and Kubo, M. (1953) *Bull. Chem. Soc. Japan,* **26,** 250: ref. 55a, M 213.

231 Vincow, G., Dauben, Jr., H.J., Hunter, F.R. and Volland, W.V. (1969) *J. Am. chem. Soc.,* **91,** 2823.

232 Rinehart, Jr. K.L., Buchholz, A.C., Van Lear, G.E. and Cantrill, H.L. (1968) *J. Am. chem. Soc.,* **90,** 2983.

232a Dashevskii, V.G., Naumov, V.A. and Zaripov, N.M. (1972) *Zh. Strukt. Khim.,* **13,** 171: (1972) *Chem. Abs.,* **76,** 98976q.

233 Krebs, A. and Schrader, B. (1967) *Annalen,* **709,** 46.

234 *Cf.,* Coburn, R.A. and Dudek, G.O. (1968) *J. phys. Chem.,* **72,** 1177.

235 Bertelli D.J. and Andrews, Jr., T.G. (1969) *J. Am. chem. Soc.,* **91,** 5280.

235a Reference 55b, M139s.

236 Oka, Y. and Hirai, M. (1968) *Nippon Kagaku Zasshi,* **89,** 589: (1968) *Chem. Abs.,* **69,** 54842.

237 Jones, W.M. and Ennis, C.L. (1969) *J. Am. chem. Soc.,* **91,** 6391.

238 Vincow, G., Morrell, M.L., Volland, W.V., Dauben, Jr. H.J. and Hunter, F.R (1965) *J. Am. chem. Soc.,* **87,** 3527.

239 Doering, W. von E. and Knox, L.H. (1954) *J. Am. chem. Soc.*, **76**, 3203.
240 Dewar, M.J.S. (1945) *Nature*, **155**, 50.
241 E.g. Roberts, J.D., and Caserio, M. (1965) *Basic Principles of Organic Chemistry*, Benjamin, New York.
242 Trotter, J. (1960) *Acta Cryst.*, **13**, 86.
243 Reference 55b, M 96s.
244 Smith, W.B. and Chiranjeevi, S. (1966) *J. phys. Chem.*, **70**, 3505.
245 Govil, G. (1967) *J. chem. Soc.*, (A) 1416.
246 *Cf.*, Bothner-By, A.A. and Naar-Colin, C. (1961) *J. Am. chem. Soc.*, **83**, 231: Acrivos, J.V. (1962) *Mol. Phys.*, **5**, 1: Gillies, D.G. (1970), personal communication.
247 Benson, R.C., Norris, C.L., Flygare, W.H. and Beak, P. (1971) *J. Am. chem. Soc.*, **93**, 5591.
248 Becker, E.D., Ziffer, H. and Charney, E. (1963) *Spec. Acta*, **19**, 1871.
249 Stevenson, P.E. (1965) *J. Mol. Spec.*, **17**, 58.
250 Wöhler, F. (1844) *Annalen*, **51**, 155.
251 Levy, M., Szwarc, M. and Throssell, J. (1954) *J. chem. Phys.*, **22**, 1904.
252 Errede, L.A. and Landrum, B.F. (1957) *J. Am. chem. Soc.*, **79**, 4952.
253 Tanaka, I. (1954) *J. chem. Soc. Japan*, **75**, 218 and 320.
253a Williams, D.J., Pearson, J.M. and Levy, M. (1970) *J. Am. chem. Soc.*, **92**, 1437.
254 Long, R.E., Sparks, R.A. and Trueblood K.N. (1965) *Acta Cryst.*, **18**, 932.
255 Melby, L.R. (1965) *Can. J. Chem.*, **43**, 1448.
256 Hoekstra, A., Spoelder, T. and Vos, A. (1972) *Acta Cryst., B*, **28**, 14.
257 Nishimoto, K. and Forster, L.S. (1966) *Theor. Chim. Acta*, **4**, 155.
258 Julg, A., Béry, J.C. and Bonnet, M. (1964) *Tetrahedron*, **20**, 2237.
259 Dewar, M.J.S. and Gleicher, G.J. (1966) *J. chem. Phys.*, **44**, 759.
260 Ellison, F.O. (1963) *J. Am. chem. Soc.* **85**, 3540: Edwards, T.G. and Grinter, R. (1968) *Theor. Chim. Acta*, **12**, 387.
261 Smitherman, H.C. and Ferguson, L.N. (1968) *Tetrahedron*, **24**, 923.
262 Mayer, R., Bray, W. and Zahradník, R. (1967) *Adv. Heterocyclic Chem.*, **8**, 219.
263 Toussaint, J. (1956) *Bull. Soc. chim. Belge*, **65**, 213: ref. 55b, M 141s.
264 Reference 67 and general texts such as 28 and 29.
265 Arndt, F. Martin, G.T.O. and Partington, J.R. (1935) *J. chem. Soc.*, 602.
266 Jonáš, J., Derbyshire, W. and Gutowsky, H.S. (1965) *J. phys. Chem.*, **69**, 1.
267 Brown, N.M.D. and Bladon, P. (1965) *Spec. Acta*, **21**, 1277 and refs. therein.
268 Mayo, R.E. and Goldstein, J.H. (1967) *Spec. Acta*, **23A**, 55.
269 Finch, P. and Gillies, D.G., personal communication.
270 Martin, J-C (1970) *Bull. Soc. chim. France*, 277.
271 Jones, R.N., Angell, C.L., Ito, T. and Smith, R.J.D. (1959) *Can. J. Chem.*, **37**, 2007.
272 Tolmachev, A.I., Schulezhko, L.M. and Kisilenko, A.A. (1968) *Zh. Obshch. Khim.*, **38**, (1) 118: (1968) *Chem. Abs.* **69**, 30591.
273 Culbertson, G. and Pettit, R. (1963) *J. Am. chem. Soc.*, **85**, 741.
274 Cavalieri, L.F. (1947) *Chem. Rev.*, **41**, 525.
275 Parkányi, C. and Zahradník, R. (1962) *Coll. Czech. Chem. Comm.*, **27**, 1355.
276 Vorlander, D. (1905) *Annalen* **341**, 64: *Cf.*, Woods, L.L. (1958) *J. Am. chem. Soc.*, **80**, 1440.
277 Jones, R.G. and Mann, M.J. (1953) *J. Am. chem. Soc.*, **75**, 4048: Ainsworth, C. and Jones, R.G. (1954) *J. Am. chem Soc.*, **76**, 3172.
278 Kehrmann, F. and Duttenhöfer, A. (1906) *Chem. Ber.*, **39**, 1299.
279 Beak, P. (1964) *Tetrahedron*, **20**, 831.
280 Zahradník, R., Párkányi, C. and Koutecký, J. (1962) *Coll. Czech. Chem. Comm.*, **27**, 1242.
281 Harris, R.L.N., Johnson, A.B. and Kay, I.T. (1966) *Q.Rev.chem.Soc.*, **20**, 240.
282 Chen. B.M.L. and Tulinsky, A. (1972) *J. Am. chem. Soc.*, **94**, 4144.
283 Silvers, S. and Tulinsky, A. (1964) *J. Am. chem. Soc.*, **86**, 927: Hamor, M.J. Hamor, T.A. and Hoard, J.L. (1964) *J. Am. chem. Soc.*, **86**, 1938: Hoard, J.L., Hamor, M.J. and Hamor, T.A. (1963) *J. Am. chem. Soc.*, **85**, 2334.
284 Fleischer, E.B. (1963) *J. Am. chem. Soc.*, **85**, 1353.
285 Koenig, D.F. (1965) *Acta Cryst.*, **18**, 663.
286 Kuhn, H. and Huber, W. (1959) *Helv. Chim. Acta*, **42**, 363.
287 Webb, L.E. and Fleischer, E.B. (1965) *J. Am. chem. Soc.*, **87**, 667.
287a Doddrell, D. and Caughey, W.S. (1972) *J. Am. chem. Soc.*, **94**, 2510.
288 Brown, C.J. (1968) *J. chem. Soc., (A)* 2488.
289 Brown, C.J. (1968) *J. chem. Soc., (A)* 2494.

290 Hoskins, B.F., Mason, S.A. and White, J.C.B. (1969) *Chem. Comm.*, 554.
291 Johnson, A.W. and Oldfield, D (1965) *J. chem. Soc.*, 4303: Abraham, R.J., Jackson, A.H., Kenner G.W. and Warburton, D. (1963) *J. chem. Soc.*, 853: Caughey, W.S. and Iber, P.K. (1963) *J. org. Chem.*, **28,** 269: Abraham, R.J. Jackson, A.H., Kenner, G.W. (1961) *J. chem. Soc.*, 3468: Caughey, W.S. and Koski, W.S. (1962) *Biochemistry*, **1,** 923.
291a Reference 35b, p. 97.
291b Broadhurst, M.J., Grigg, R. and Johnson, A.W. (1972) *J. chem. Soc., Perkin 1,* 2111.
292 Eisner, U. and Linstead, R.P. (1955) *J. chem. Soc.*, 3749: Stern, E.S. and Timmons, C.J. (1970) *'Gillam and Stern's Introduction to Electronic Absorption Spectroscopy in Organic Chemistry'* p. 176. E. Arnold Ltd., London.
293 Whalley, M. (1961) *J. chem. Soc.*, 866.
294 Sundbom, M. (1968) *Acta chem. scand.* **22,** 1317.
295 Samuels, E., Shuttleworth, R. and Stevens, T.S. (1968) *J. chem. Soc.*, (C) 145.
296 Inhoffen, H.H., Fuhrhop, J-H., Voeight, H. and Brockmann, Jr., H. (1966) *Annalen,* **695,** 133.
297 Winstein, S., ref. 11, p. 5: Winstein, S. (1972) in *'Carbonium Ions'*, ed. Olah, G.A. and von Schleyer, P., Wiley-Interscience, Vol. 3, Ch. 22: Winstein, S. and Sonnenberg, J. (1961) *J. Am. chem Soc.*, **83,** 3244.
298 Cowling, S.A., Johnstone, R.A.W., Gorman, A.A. and Smith, P.G. (1973) *J. chem. Soc.. Chem. Comm..* 627.
299 Diaz, A., Brookhart, M. and Winstein, S., (1966) *J. Am. chem. Soc.*, **88,** 3133.
300 Sakai, M. (1974) *J. chem. Soc., Chem. Comm.*, 6.
301 Brown, H.C. (1966) *Chemistry in Britain,* **2,** 199.
302 Olah, G.A. and White, A.M. (1969) *J. Am. chem. Soc.*, **91,** 6883.
303 Peters, D. (1963) *J. chem. Soc.*, 2015 *et seq.*
304 Hehre, W.J. (1973) *J. Am. chem. Soc.*, **95,** 5807.
305 Winstein, S., Moshuk, G., Rieke, R. and Ogliaruso, M. (1973) *J. Am. chem. Soc.*, **95,** 2624.
306 Berson, J.A. and Jenkins, J.A. (1972) *J. Am. chem. Soc.*, **94,** 8907.
307 Winstein, S. Ogliaruso, M., Sakai, M. and Nicholson, J.M. (1967) *J. Am. chem. Soc.*, **89,** 3656.
308 Breslow, R., Brown, J. and Gajewski, J.J. (1967) *J. Am. chem. Soc.*, **89,** 4383.
309 Dewar, M.J.S., Reference 11, p. 206 *et seq.*
310 Breslow, R. and Mazur, S. (1973) *J. Am. chem. Soc.*, **95,** 584.
311 Saunders, M., Berger, R., Jaffe, A., McBride, J.M., O'Neill, J., Breslow, R., Hoffman, J.M. Jr., Perchonock, C., Wasserman, E., Hutton, R.S., Kuck, V.J. (1973) *J. Am. chem. Soc.*, **95,** 3017.
312 Ha, T-K., Graf, F., and Günthard, H.H. (1973) *J. Mol. Structure,* **15,** 335.
313 Chapman, O.L., McIntosh, C.L. and Pacansky, J. (1973) *J. Am. chem. Soc.*, **95,** 614.
314 Krantz, A., Lin, C.Y. and Newton, M.D. (1973) *J. Am. chem. Soc.*, **95,** 2744.
315 Warner, P. and Winstein, S. (1969) *J. Am. chem. Soc.*, **91,** 7785.
316 Staley, S.W. and Pearl, N.J. (1973) *J. Am. chem. Soc.*, **95,** 2731.

Index

acid-base equilibria 22
acidity, hydrocarbons 22
aliphatic compounds 2
annulenes
  bond length alternation 27,55
  p.m.r. 27
[10] annulene 63
  1,6-imino- 63
  1,6-methano- 63,92
  1,6-methano- 2-carboxylic acid, molecular geometry 63
  1,6-oxido- 63
[16]annulene
  bond length alternation 64
  diamagnetic susceptibility exaltation 29
[18] annulene
  bond length computations 64
  chemistry 64
  diamagnetic anisotropy 64
  diamagnetic susceptibility exaltation 64
  electronic spectrum 64
  molecular geometry 63
  p.m.r. 26,64
  resonance energy 64
antiaromaticity 90
  charge distribution 91
  definition 90,92
aromatic character
  bond length criterion 3,11,79
  bond order 14
  chemical reactivity criterion 4,23,94
  closed shell of electrons 30
  Craig's definition 6,68
  definition 3,4,5,31,86,94
  evaluation from wave function 43
  excited electronic states 7,62,95
  'from the centre' 53
  historical background 1
  magnetic criteria 4
  molecular geometry 10
  new areas 7
  planarity of ring 6,12
  role of π and σ-electrons 6
  substituent effects 24
  thermodynamic criterion 3
  transition states 94
aromatic molecules, strained 22
aromatic sextet 2,30,43,85
aromaticity, bicyclo- 6
aromaticity constant 11
  numerical values 12
aromaticity, excited electronic states 7,62,92
atomisation energy 15
Aurbau principle 7

azulene
  analogues 62
  bond length computations 61
  chemistry 62
  diamagnetic anisotropy 62
  diamagnetic susceptibility exaltation 62
  dipole moment 62
  1,3-dipropionic acid, molecular geometry 61
  excited electronic states 62,95
  molecular orbital energy levels 48
  p.m.r. 62,74
  resonance energy 61

benzene 94
  bond lengths 11
  diamagnetic anisotropy 29
  diamagnetic susceptibility exaltation 29
  excited electronic states 7
  heat of formation 14
  heat of hydrogenation 21
  molecular orbitals 42
  molecular orbital energies 57
  molecular orbital energies and electron populations 46
  molecular orbital energy levels 48
  π-orbital energy levels 57
  p.m.r. 25
  radical anion 30
  radical cation 31
  resonance energy 19,55
  stability 48
  $^3B_{1u}$, $^1B_{2u}$ states 7
p-benzoquinone
  chemical reactivity 77
  computations 78
  diamagnetic anisotropy 77
  diamagnetic susceptibility exaltation 77
  electronic spectrum 77
  infrared spectrum 77
  molecular geometry 76
  p.m.r. 76
  resonance energy 76,78
benzyne 8

cis-3-bicyclo [3,1,0] hexyl tosylate 87
bicyclo [3,3,0] octa-3,7-diene-2,6-dione 68
bicyclo [3,2,1] octadienyl anion 89
bitropenyl 75
bond energy 15,51
  calculations 18
  Franklin's method 59

Klages' method 19,59
bond lengths 10
coupling constants 28
determination 12
excited electronic state 7
bonds, localised 5
bond order 14,61
coupling constants 28
bond polarity 46
Born Oppenheimer approximation 43
butadiene, bond lengths 11

carbocycles, seven-membered 71
carbon monoxide, dipole moment 34
charge transfer 46
chemical bond 2
two electron 2,40
chemical reactivity – see also under individual compounds
geometry 24
chemical valence theory 40
α-chlorohaemin, molecular geometry 82
chlorophylls 81
CNDO theory 54
complexes, benzene 18
computations, accurate, on large molecules 37
coupling constants 14
^{13}C-H 28
^{3}J 4,27
cross conjugation 2,67
cyclobutadiene 92
cycloheptatriene – see under tropylidene
cycloheptatrienylidene 75
cyclohexatriene 18,21
cyclooctatetraene
ionisation potential 34
p.m.r. 26
cyclooctatrienyl anion, 8,8-dimethyl 93
cyclopentadiene
bond lengths 11
diamagnetic anisotropy 30
dipole moment 35
pKa 22
cyclopentadienyl anion 22
molecular orbital energy levels 48,57
p.m.r. 26
cyclopentadienyl cation 90
cyclopenta [*b*] pyran 63
cyclopenta [*c*] thiapyran 62
cyclopropenone, dialkyl and diaryl, dipole moments 35
cyclopropenyl anion 90
molecular geometry 90
π-orbital energies 90
cyclopropenyl cation 6,86,90
π-orbital energies 90
triphenyl, bond lengths 13
triphenyl, pK_a 22
cytochromes 81

diamagnetic anisotropy – see also under individual compounds 4,28
numerical values 29
diamagnetic susceptibility 28,30
numerical values 29
diamagnetic susceptibility exaltation – see also under individual compounds 30
numerical values 29
dipole moments
aromatic character and valence bond structures 34

eigenvalue 50
electron affinity 35
electron delocalisation 90
electronegativity 47
electronic spectra – see also under individual compounds
electronic spectroscopy 32
e.s.r. 30,75,89
electron delocalisation 30
eucarvone
pK_a 74
p.m.r. 73
excitation energy 50
excited electronic state 7
bond length 7
expectation values 45

Faraday effect 26
ferrocene
chemistry 64
computations 65
diamagnetic anisotropy 65
diamagnetic susceptibility exaltation 65
heat of combustion 65
infrared spectrum 65
molecular geometry 64
p.m.r. 65
resonance energy 65
σ and π-electrons 65
formaldehyde, wave function 51
Franklin's method for heats of combustion 59
free energy of activation 23
fulvene 2
chemistry 67
computations 67
diamagnetic susceptibility exaltation 67
6,6-dimethyl 11
6,6-dimethyl, molecular geometry 67
6,6-dimethyl, resonance energy 67
dipole moment 34,67
molecular orbital energy levels 48
p.m.r. 67

geometry, chemical reactivity 24
guanidinium cation, orbital energy levels 53

haemin, molecular geometry 82
haems 81
Hartree-Fock equation and operator 39,43
heat of atomisation 14
numerical values for hydrocarbons 16

heat of combustion 15
Franklin's method 59
numerical values for hydrocarbons 16
heat of formation 14
numerical values for hydrocarbons 16
heat of hydrogenation 15,21
numerical values 20
heat of sublimation, numerical values of hydrocarbons 16,59
heptafulvene, resonance energy 73
heterocyclic compounds, combustion 21
homoaromaticity 86
homoconjugation 86
homotropylium cation 89
Hückel theory 50,57
Hückel's rule 2,6,58,61,90
hybridisation 46
hydrocarbons, acidity 22
4-hydroxypyrylium cation 80
hyperconjugation – see 'sigmaconjugation'

1,6-imino[10]annulene 63
infrared spectroscopy 31
infrared spectra – see under individual compounds
ionisation potential 32,50
ions, hydrogenation 22

Kekulé structures 8
Klages' method for bond energies 19,59
Koopman's theorem 33

mass spectroscopy 35
methane
electron populations 46
molecular orbitals 33,41
1,6-methano[10]annulene
2-carboxylic acid, molecular geometry 63,92 63
diamagnetic susceptibility exaltation 29
2-methylenebicyclo[3,2,1]octa-3,6-diene 87
5-methylenenorborn-2-ene 86
molecular energy, total 49,51
molecular geometry 7,88
bond lengths 10
molecular orbital theory 2
CNDO 54
Huckel theory 50,57
non-empirical versus semi-empirical 50
qualitative ideas 38
self-consistent field 43,45
self-consistent approximation 49
σ and π-electron separation 49
σ-electrons as simple bonds 55
molecular orbitals
canonical form 33,40
eigenvalue equation 45
lack of uniqueness 32,41,91
linear transformation 41,91
localised 40,49
zero energy line 54
monohomobenzene 89
monohomocyclopropenyl cation 86

naphthalene 11
bondlengths, molecular orbital energy levels 48
non-aromatic compounds 5
non-benzenoid compounds 3
non-classical ions 86
n.m.r. 88
non-classical ions versus equilibrating classical ions 88
norbornadiene 86
norbornenyl cation 87
anti-7-norbornenyl tosylate, solvolysis 87
norbornyl cation 88
7-norbornyl tosylate, solvolysis 87
nuclear magnetic resonance 25
nuclear quadrupole moment 35

ω-technique 49,57
operator, Hamiltonian 43
1,6-oxido[10]annulene 63

Pariser-Parr-Pople method – see CNDO Theory
pentalene 6
1,3-bis (dimethylamino)- 68
computations 68
derivatives 68
dianion 68
electronic spectrum 68
preparation 68
vibrational decomposition 68
phenanthrene
bond length computations 60
chemistry 60
diamagnetic anisotropy 60
diamagnetic susceptibility exaltation 60
molecular geometry 60
p.m.r. 60
resonance energy 60
photoelectron spectroscopy 32
phthalocyanin 81
diamagnetic anisotropy 84
molecular geometries 83
π-electron, definition 37
polarography 35
populations, electron 46
porphin 81
^{13}C.m.r. 83
chemical reactivity 84
electronic spectra 84
'great ring' 83
and its derivatives, molecular geometries 82
resonance energy 8,84
porphyrins 81
chemical reactivity 84
p.m.r. 84
properties, one electron 38
p.m.r. 25
^{3}J coupling constants, numerical values 28
shift values 26

pyran-4-thione-2,6-dimethyl, molecular
 geometry 79
pyridine, molecular orbital energies
 and electron populations 47
2-pyrindine 63
2-pyrone, 4-methoxy-6-methyl- 81
4-pyrone
 chemistry 80
 computations 81
 diamagnetic anisotropy 79
 2,6-dimethyl-, diamagnetic
 susceptibility exaltation 79
 infrared spectrum 80
 pK_a 80
 p.m.r. 79
 resonance energy 79
pyrrole, molecular orbital energies
 and electron populations 47
pyrylium cation 81
 4 hydroxy- 80
p-quinodimethane
 electronic spectrum 77
 p.m.r. 77

quinoid structures 76
 lack of aromaticity 78
 6π-electron ring 76

radicals, aromaticity 30
Raman spectroscopy 31
resonance energy – see also under
 individual compounds 21,50,52
 calculations 19
 numerical values 55
 reference molecule 53
ring current 4,62,84
 proton shifts 25

sapphyrin, pentaethyl, pentamethyl,
 p.m.r. 84
Schrödinger's equation 39
 rigorous solution 43
secular determinant and equations 40
selenophen 66
separation of σ and π-electrons 44,48,92
SI units, conversion factors 96
σ and π-electron interaction 51,67
σ and π-electrons, separation 44,48,92
sigmaconjugation 18,53,64
spiroconjugation 6
stabilisation energy – see
 'resonance energy' 21
standard state 14

tetracyano-*p*-quinodimethane
 molecular geometry 77
 radical anion, molecular geometry 78
 resonance energy 78
thermochemical data, numerical values 59
thermochemistry 14
thiophen
 chemistry 66
 computations 66
 derivatives, e.s.r. 66
 diamagnetic anisotropy 66
 diamagnetic susceptibility exaltation 66
 dipole moment 66
 molecular geometry 65
 p.m.r. 66
 resonance energy 66
thiopyran-4-one, 2,6-diphenyl, resonance
 energy 79
thiopyrones, p.m.r. 79
β-thujaplicin, molecular geometry 72
transannular bond 61
transition state 6,24
2-tricyclo[4,1,0,$0^{3,7}$] heptyl methyl
 ether 88
trishomocyclopropenyl cation 87
tropenyl radical 75
tropolone
 basicity 74
 p-chlorobenzoate, molecular
 geometry 72
 dipole moment 75
 4-isopropyl, molecular geometry 72
 molecular geometry 72
 pK_a 75
 p.m.r. 74
 and its derivatives, resonance energies 73
 tropolonyl *p*-chlorobenzoate,
 molecular geometry 72
tropone,
 3 azido, molecular geometry 72
 basicity 22,74
 4,5-benzoderivative, molecular
 geometry 72
 2-chloro, molecular geometry 72
 2-chloro, p.m.r. 73
 diamagnetic susceptibility exaltation 74
 and its derivatives, dipole moments 75
 infrared spectrum 74
 molecular geometry 72
 pK_a 74
 and its derivatives, p.m.r. 74
 and its derivatives, resonance energies 72
tropylidene
 diamagnetic susceptibility exaltation 89
 molecular geometry 73
 resonance energy 73
tropylium cation 24
 bond lengths 13
 12-dihydroxy, molecular geometry 75
 1,2-dihydroxy, p.m.r. 75
 hydroxy 74
 hydroxy-, p.m.r. 74
 infrared spectrum 31
 molecular geometry 72
 π-orbital energy levels 57
 p.m.r. 27
 resonance energy 72

ultraviolet spectroscopy – see 'electronic
 spectroscopy'

valence bond theory 2,78
vibration spectroscopy 31

wave function 38
  antisymmetric 41
  computations 'molecules in molecules' 78
  configuration interaction 44
  exact 43
Wheland-Mann method 49,57

p-xylylene– see under *p*-quinodimethane

zero differential overlap 54